PLL Performance, Simulation, and Design
3rd Edition

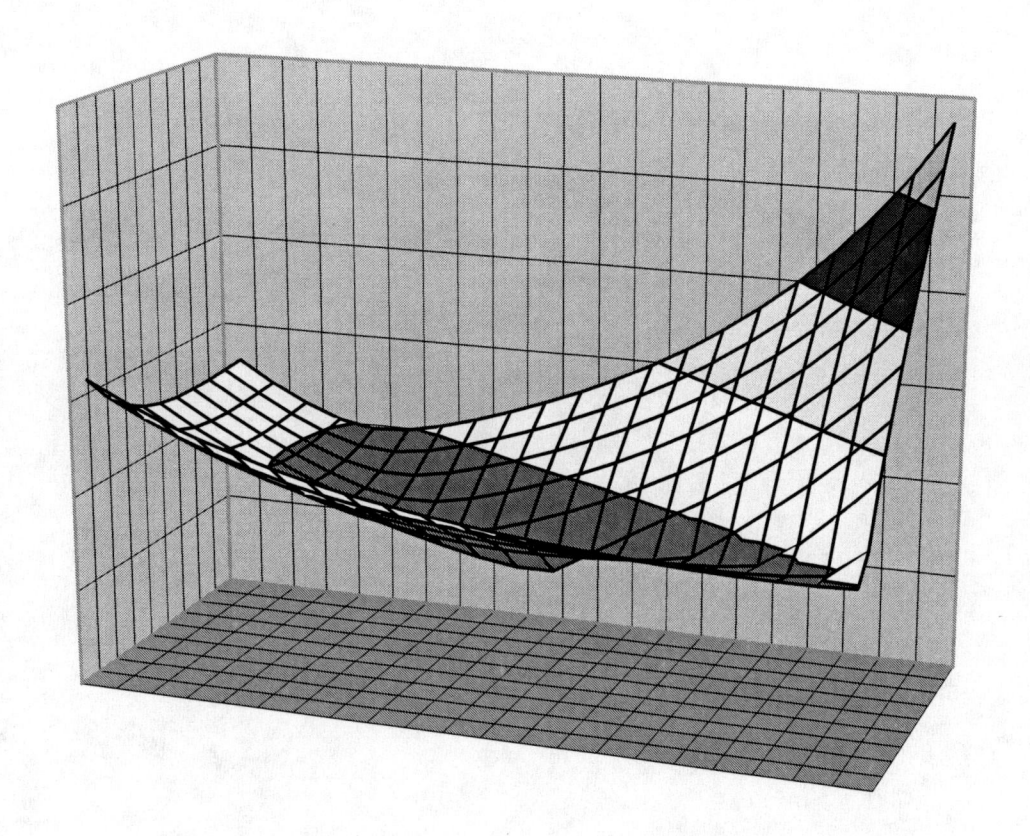

Dean Banerjee

To Caleb

Credits

I would like to thank the following people for their assistance in making this book possible. Some of these people helped directly with things like editing and cover design, while others have helped in indirect ways like useful everyday conversation and creating things that helped me grow in my understanding of PLLs.

Person	Editing 2nd Edition	3rd Edition	Useful Insights
Yuko Kanagy	-	-	Useful insights into PLLs
Bill Keese	-	-	Wrote National Semiconductor Application Note 1001, which was my first introduction to loop filter design.
Tom Mathews	-	-	Useful insights into RF phenomena.
Khang Nguyen	-	-	Developed the GUI for EasyPLL at wireless.national.com that is based on many of the formulas in this book.
Ian Thompson	-	-	Useful insights into PLLs, particularly phase noise and how it is impacted by the discrete sampling action of the phase detector.
Timothy Toroni	-	-	Developed a TCL interface for many of my simulation routines in C that proved to be very useful.
Deborah Brown	X	-	Thorough editing from cover to cover
Bill Burdette	X	-	
Stephen Hoffman	-	X	
Shigura Matsuda	X	-	Translation into Japanese.
John Johnson	X	X	Special thanks to John Johnson for doing the cover design for the 3rd Edition and thorough editing from cover to cover. John also did a lot of the illustrations
Tien Pham	-	X	Cover to cover editing.
Ahmed Salem	-	X	
Benyoung Zhang	-	X	Useful insights into delta-sigma PLLs in general and the LMX2470 in particular.

Preface

I first became familiar with PLLs by working for National Semiconductor as an applications engineer. While supporting customers, I noticed that there were many repeat questions. Instead of creating the same response over and over, it made more sense to create a document, worksheet, or program to address these recurring questions in greater detail and just re-send the file. From all of these documents, worksheets, and programs, this book was born.

Many questions concerning PLLs can be answered through a greater understanding of the problem and the mathematics involved. By approaching problems in a rigorous mathematical way one gains a greater level of understanding, a greater level of satisfaction, and the ability to apply the concepts learned to other problems.

Many of the formulas that are commonly used for PLL design and simulation contain gross approximations with no or little justification of how they were derived. Others are rigorously derived, but from outdated textbooks that make assumptions not true of the PLL systems today. It is therefore no surprise that there are so many rules of thumb to be born which yield unreliable results. Another fault of these formulas is that many of them have not been compared to measured data to ensure that they account for all relevant factors.

There is also the other approach, not trusting formulas enough and relying on only measured results. The fault with this is that many great insights are lost and it is difficult to learn and grow in PLL knowledge this way. Furthermore, by knowing what a result should theoretically be, it makes it easier to spot and diagnose problems with a PLL circuit. This book takes a unique approach to PLL design by combining rigorous mathematical derivations for formulas with actual measured data. When there is agreement between these two, then one can feel much more confident with the results.

The purpose of writing a third edition is to add significant details and understanding to what was in the second edition. This includes insights into delta sigma PLLs, fractional spurs, phase noise, lock time, and loop filter design.

Table of Contents

PLL BASICS 9
- CHAPTER 1 BASIC PLL OVERVIEW 11
- CHAPTER 2 THE CHARGE PUMP PLL WITH A PASSIVE LOOP FILTER 13
- CHAPTER 3 PHASE/FREQUENCY DETECTOR THEORETICAL OPERATION 15
- CHAPTER 4 BASIC PRESCALER OPERATION 21
- CHAPTER 5 FUNDAMENTALS OF FRACTIONAL N PLLs 25
- CHAPTER 6 DELTA SIGMA FRACTIONAL N PLLs 30
- CHAPTER 7 THE PLL AS VIEWED FROM A SYSTEM LEVEL 34

PLL PERFORMANCE AND SIMULATION 39
- CHAPTER 8 INTRODUCTION TO LOOP FILTER COEFFICIENTS 41
- CHAPTER 9 INTRODUCTION TO PLL TRANSFER FUNCTIONS AND NOTATION 46
- CHAPTER 10 REFERENCE SPURS AND THEIR CAUSES 53
- CHAPTER 11 FRACTIONAL SPURS AND THEIR CAUSES 67
- CHAPTER 12 ON NON-REFERENCE SPURS AND THEIR CAUSES 80
- CHAPTER 13 PLL PHASE NOISE MODELING AND BEHAVIOR 88
- CHAPTER 14 RMS PHASE ERROR AND DERIVED NOISE QUANTITIES 99
- CHAPTER 15 TRANSIENT RESPONSE OF PLL FREQUENCY SYNTHESIZERS 106
- CHAPTER 16 DISCRETE LOCK TIME ANALYSIS 120
- CHAPTER 17 ROUTH STABILITY FOR PLL LOOP FILTERS 127
- CHAPTER 18 A SAMPLE PLL ANALYSIS 132

PLL DESIGN 145
- CHAPTER 19 FUNDAMENTALS OF PLL PASSIVE LOOP FILTER DESIGN 147
- CHAPTER 20 EQUATIONS FOR A PASSIVE SECOND ORDER LOOP FILTER 151
- CHAPTER 21 EQUATIONS FOR A PASSIVE THIRD ORDER LOOP FILTER 155
- CHAPTER 22 EQUATIONS FOR A PASSIVE FOURTH ORDER LOOP FILTER 163
- CHAPTER 23 FUNDAMENTALS OF PLL ACTIVE LOOP FILTER DESIGN 173
- CHAPTER 24 ACTIVE LOOP FILTER USING THE DIFFERENTIAL PHASE DETECTOR OUTPUTS 184
- CHAPTER 25 IMPACT OF LOOP FILTER PARAMETERS AND FILTER ORDER ON REFERENCE SPURS 187
- CHAPTER 26 OPTIMAL CHOICES FOR PHASE MARGIN AND GAMMA OPTIMIZATION PARAMETER 195
- CHAPTER 27 DEALING WITH REAL-WORLD COMPONENTS 213
- CHAPTER 28 USING FASTLOCK AND CYCLE SLIP REDUCTION 202
- CHAPTER 29 SWITCHED AND MULTIMODE LOOP FILTER DESIGN 209

ADDITIONAL TOPICS 217
- CHAPTER 30 LOCK DETECT CIRCUIT CONSTRUCTION AND ANALYSIS 219
- CHAPTER 31 IMPEDANCE MATCHING ISSUES AND TECHNIQUES FOR PLLs 226
- CHAPTER 32 OTHER PLL DESIGN AND PERFORMANCE ISSUES 233

SUPPLEMENTAL INFORMATION 243
- CHAPTER 33 GLOSSARY AND ABBREVIATION LIST 245
- CHAPTER 34 REFERENCES 254
- CHAPTER 35 USEFUL WEBSITES AND ONLINE RF TOOLS 255

PLL Basics

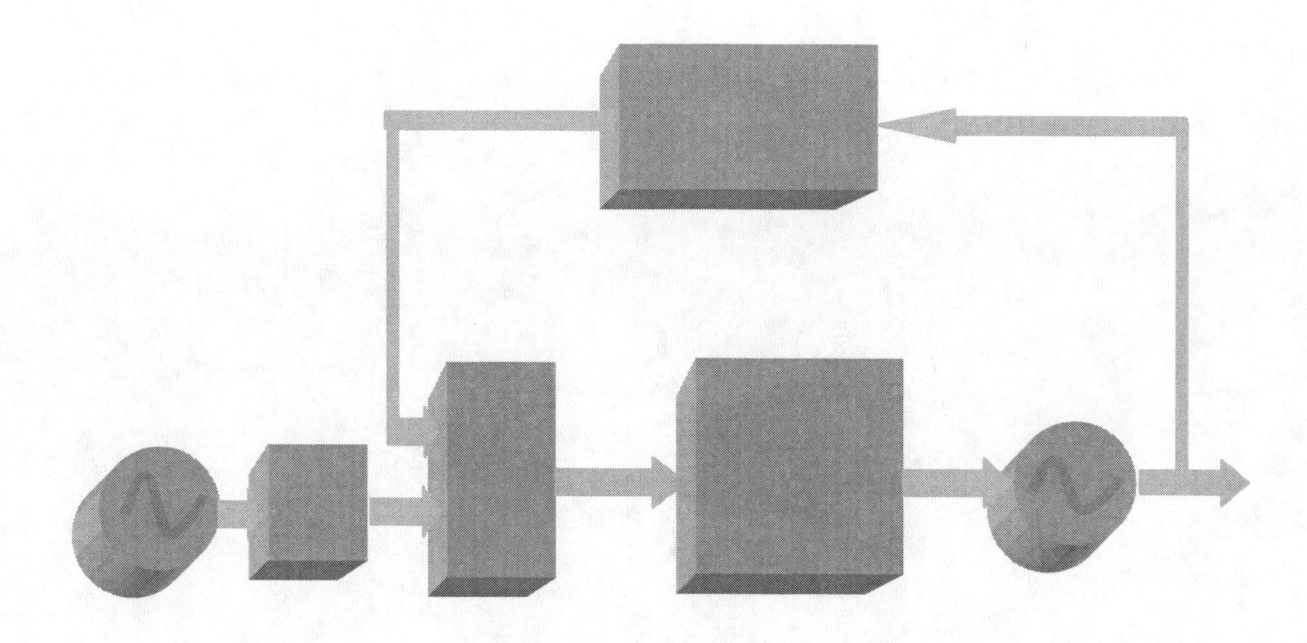

Chapter 1   Basic PLL Overview

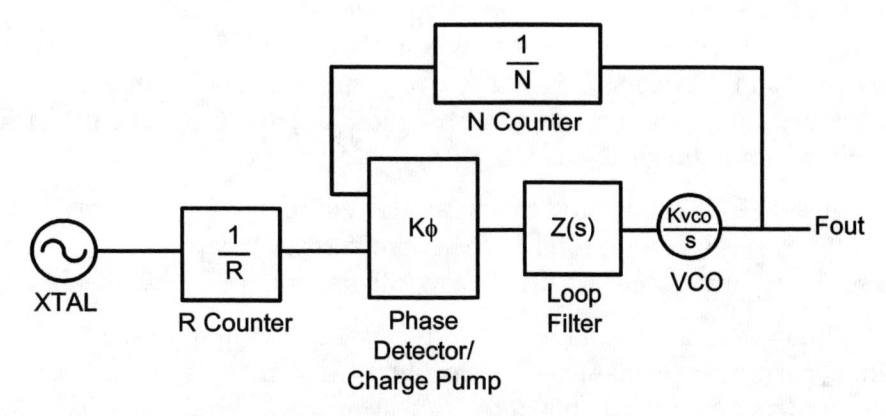

Figure 1.1 *The Basic PLL*

Basic PLL Operation and Terminology
This section describes basic **PLL** (Phase-Locked Loop) operation and introduces terminology that will be used throughout this book. The PLL starts with a stable crystal reference frequency, **XTAL**, which is divided down to a lower frequency by the **R counter**. This divided frequency is called the comparison frequency (**Fcomp**) and is one of the inputs to the phase detector. The phase-frequency detector outputs a current that has an average DC value proportional to the phase error between the comparison frequency and the output frequency, after it is divided by the **N** divider. The constant of proportionality is called **Kϕ**. This constant turns out to be the magnitude of the current that the charge pump can source or sink. Although it is technically correct to divide this term by 2π, it is unnecessary since it is canceled out by another factor of 2π which comes from the VCO gain for all of the equations in this book. So technically, the units of **Kϕ** are expressed in mA/(2π radians).

If one takes this average DC current value from the phase detector and multiplies it by the impedance of the loop filter, **Z(s)**, then the input voltage to the **VCO** (Voltage Controlled Oscillator) can be found. The VCO is a voltage to frequency converter and has a proportionality constant of **Kvco**. The loop filter is a low pass filter, often implemented with discrete components. The loop filter is application specific, and much of this book is devoted to the loop filter. This tuning voltage adjusts the output phase of the VCO, such that its phase, when divided by *N*, is equal to the phase of the comparison frequency. Since phase is the integral of frequency, this implies that the frequencies will also be matched, and the output frequency will be given by:

$$Fout = \frac{N}{R} \cdot XTAL \qquad (1.1)$$

This applies only when the PLL is in the locked state; this does not apply during the time when the PLL is acquiring a new frequency. For a given application, **R** is typically fixed, and the *N* value can easily be changed. If one assumes that *N* and *R* must be an integer, then this implies that the PLL can only generate frequencies that are a multiple of **Fcomp**. For this reason, many people think that *Fcomp* and the channel spacing are the same. Although this is often the case, this is not necessarily true. For a fractional N PLL, *N* is not restricted to an integer, and therefore the comparison frequency can be chosen to be much larger than the channel spacing. There are also less common cases in which the comparison frequency is chosen smaller than the channel spacing to overcome restrictions on the allowable values of *N*, due to the prescaler. In general, it is preferable to have the comparison frequency as high as possible for optimum performance.

Note that the term PLL technically refers to the entire system shown in Figure 1.1 ; however, sometimes it is meant to refer to the entire system except for the crystal and VCO. This is because these components are difficult to integrate on a PLL synthesizer chip.

The transfer function from the output of the R counter to the output of the VCO determines a lot of the critical performance characteristics of the PLL. The closed loop bandwidth of this system is referred to as the loop bandwidth (*Fc*), which is an important parameter for both the design of the loop filter and the performance of the PLL. Note that *Fc* will be used to refer to the loop bandwidth in Hz and ωc will be used to refer to the loop bandwidth in radians. Another parameter, phase margin (ϕ), refers to 180 degrees minus the phase of the open loop phase transfer function from the output of the R counter to the output of the VCO. The phase margin is evaluated at the frequency that is equal to the loop bandwidth. This parameter has less of an impact on performance than the loop bandwidth, but still does have a significant impact and is a measure of the stability of the system.

The PLL as a Frequency Synthesizer

The PLL has been around for many decades. Some of its earlier applications included keeping power generators in phase and synchronizing to the sync pulse in a TV Set. Still other applications include recovering a clock from asynchronous data and demodulating an FM modulated signal. However, the focus of this book is the use of a PLL as a frequency synthesizer.

In this type of application, the PLL is used to generate a set of discrete frequencies. A good example of this is FM radio. In FM radio, the valid stations range from 88 to 108 MHz, and are spaced 0.1 MHz apart. The PLL generates a frequency that is 10.7 MHz less than the desired channel, since the received signal is mixed with the PLL signal to always generate an IF (Intermediate Frequency) of 10.7 MHz. Therefore, the PLL generates frequencies ranging from 77.3 MHz to 97.3 MHz. The channel spacing would be equal to the comparison frequency, which would is 100 kHz.

A fixed crystal frequency of 10 MHz can be divided by an *R* value of 100 to yield a comparison frequency of 100 kHz. Then the *N* value ranging from 773 to 973 is programmed into the PLL. If the user is listening to a station at 99.3 MHz and decides to change the channel to 103.4 MHz, then the *R* value remains at 100, but the *N* value changes from 886 to 927. The performance of the radio will be impacted by the spectral purity of the PLL signal produced and also the time it takes for the PLL to switch frequencies.

The loop filter has a large impact on how long it takes for the PLL to switch frequencies and also on how spectrally pure the PLL signal produced is. For this reason, there is a big emphasis on loop filter design in this book.

Chapter 2   The Charge Pump PLL with a Passive Loop Filter

Introduction

The phase detector is a device that converts the differences in the two phases from the N counter and the R counter into an output voltage. Depending on the technology, this output voltage can either be applied directly to the loop filter, or converted to a current by the charge pump.

The Voltage Phase Detector Without a Charge Pump

This type of phase detector outputs a voltage directly to the loop filter. There are several ways that it can be implemented. Possible implementations include a mixer, XOR gate, or JK Flip Flop. In the case of all these implementations there are some limitations. If the loop filter is passive, the PLL can not lock to the correct frequency if target frequency or phase is too far off from that of the VCO. Also, once the PLL is in lock, it can fall out of lock if the VCO signal goes more than a certain amount off in freqeuncy. Even when the PLL is in lock, there is steady state phase error. For instance, the mixer phase detector introduces a 90 degree phase shift. There are ways around these problems such as using acquisition aids or using active filters. Although active filters do fix a lot of these problems, op-amps add cost and noise. Floyd Gardner's classical book, *Phaselock Techniques*, goes into great detail about all the details and pitfalls of this sort of phase detector. Gardner's book presents the following topology for active loop filters.

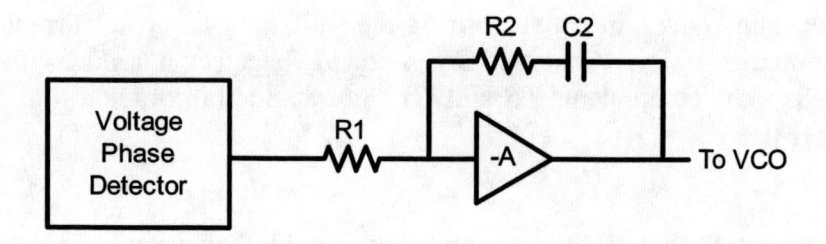

Figure 2.1 *Classical Active Loop Filter Topology for a Voltage Phase Detector*

The Modern Phase Frequency Detector with Charge Pump and its Advantages

The phase/frequency detector (**PFD**) does a much better job dealing with a large error in frequency. It is typically accompanied with a charge pump. The PFD converts the phase error presented to it into a voltage, which is in turn converted in to a correction current by the charge pump. Because these two devices are typically integrated together on the same chip and work together, the terminology is often misused. The term of PFD can be used to refer to the device that only converts the error phase into a voltage, or also can be used to refer to the device with the charge pump integrated with it. The term of charge pump is only used to refer to the device that converts the error voltage to a correction current. However, it is understood that a charge pump PLL also has a phase/frequency detector, because a charge pump is always used with a phase frequency detector. Even though the PFD and charge pump are technically separate entities, the terms are often interchanged.

Now that the use and abuse of the terminology has been discussed, it is time to discuss the benefits of using these devices. The charge pump PLL offers several advantages over the voltage phase detector and has all but replaced it. Using the PFD, the PLL is able to lock to any frequency, regardless of how far off it initially is in frequency and does not have a steady state phase error. The PFD shown in Figure 2.2 can be compared to its predecessor in Figure 2.1 .

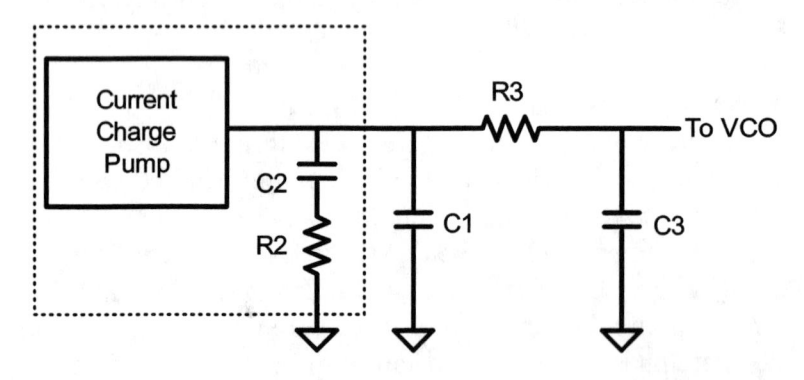

Figure 2.2 *Passive Loop Filter with PFD*

The functionality of the classical voltage phase detector and op-amp is achieved with the charge pump as shown inside the dotted lines. It is necessary to divide the voltage phase detector voltage gain by **R1** in order convert the voltage gain to a current gain for the purposes of comparison. The capacitor **C1** is added, because it reduces the spur levels significantly. Also, the components **R3** and **C3** can be added in order to further reduce the reference spur levels.

Conclusion
The classical voltage phase detector was the original implementation used for PLLs. There is excellent literature covering this device, and it is also becoming outdated. The charge pump PLL, which is the more modern type of PLL, has a phase/frequency detector and charge pump that overcomes many of the problems of its predecessor. Although op-amps can be used with the voltage phase detector to overcome many of the problems, the op-amp adds cost, noise, and size to a design, and is therefore undesirable. The only case where the op-amp is really necessary is when the VCO tuning voltage needs to be higher than the charge pump can supply. In this case, an active filter is necessary. The focus on this book is primarily on charge pump PLLs because this technology is more current and the fact that there is already a substantial amount of excellent literature on the older technology.

Chapter 3 Phase/Frequency Detector Theoretical Operation

Introduction

Perhaps the most difficult component to understand in the PLL system is the phase/frequency detector (PFD). Technically, the phase/frequency detector converts the phase error from these two signals into a voltage, which is in turn converted to a current by the charge pump. Since phase is the integral of frequency, it also gives some indication of the frequency error.

Looking carefully at Figure 3.1, it should be clear that the output is modeled as a phase and not a frequency. The VCO gain is divided by s, which corresponds to integration. The reason for this is that phase is the integral of frequency. If the frequency output is sought, then it is only necessary to multiply the transfer function by a factor of s, which corresponds to differentiation. Because phase is the integral of frequency, it follows that the PFD detects differences in both phase and frequency.

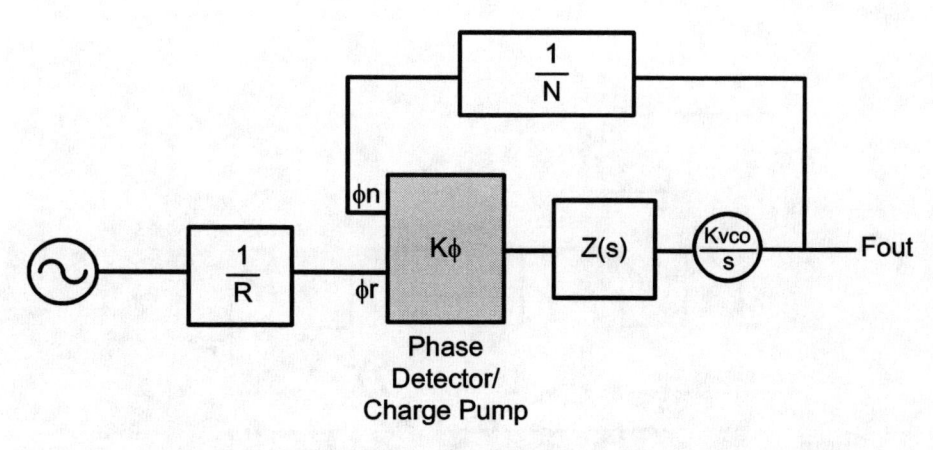

Figure 3.1 *The Basic PLL Structure Showing the Phase/Frequency Detector*

Analysis of the Phase/Frequency Detector

The output phase of the VCO is divided by N, before it gets to the Phase-Frequency Detector (**PFD**). Let ϕn represent the phase of this signal at the PFD, and fn represent the frequency of this signal. The output phase of the crystal reference is divided by R before it gets to the PFD. Let ϕr be the phase of this signal and fr be the frequency of this signal. The PFD is only sensitive to the rising edges of ϕr and ϕn.

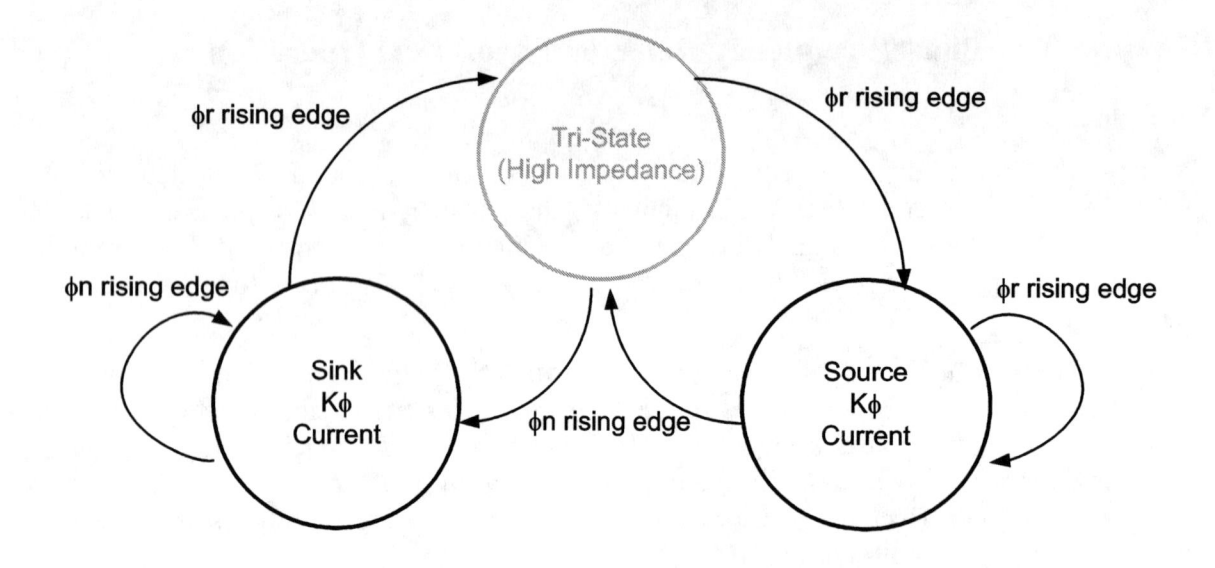

Figure 3.2 *States of the Phase Frequency Detector (PFD)*

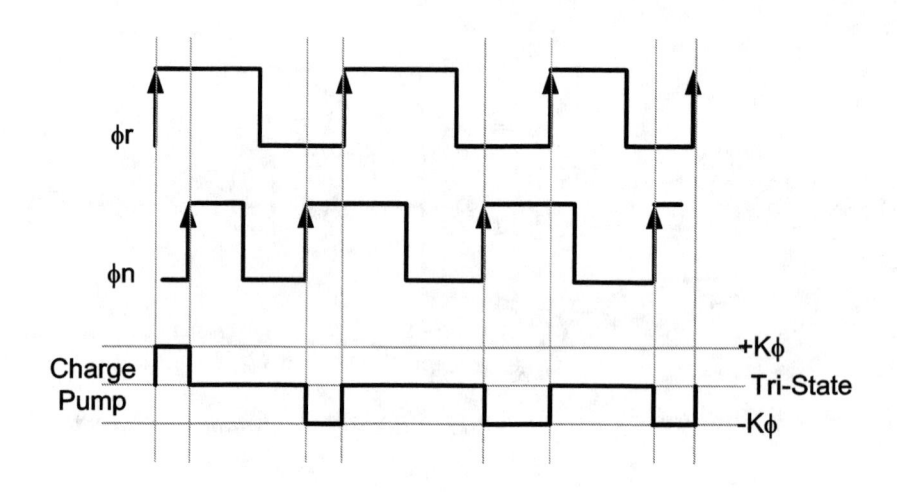

Figure 3.3 *Example of how the PFD works*

The PFD is only sensitive to the rising edges of these signals. Figure 3.2 and Figure 3.3 demonstrate its operation. Whenever there is a rising edge from the output of the R counter (shown by the symbol ϕr), there is the positive transition from the charge pump. This means that if the charge pump was sinking current, then it now is in a Tri-State mode. If it was in Tri-State, then it is now sourcing current. If it already was sourcing current, then it continues to source current. The rising edges from the N counter work in an analogous way, except that it causes negative transitions for the charge pump. If the charge pump was sourcing current, it now goes to Tri-State. If it was in Tri-State, it now goes to sourcing current, if it was sourcing current, it continues to source current.

Analysis of the PFD for a Phase Error

Suppose that ϕn and ϕr are at the exact same frequency but off in phase such that the leading edge of ϕr is leading the leading edge of ϕn by a constant time period equal to τ. There are two cases that need to be covered.

$\tau = 0$: For this case, there is no phase error, and the signals are synchronized in frequency and phase, therefore there would theoretically be no output of the phase detector. In actuality, there would be some very small outputs from the phase detector due to dead zone elimination circuitry and gate delays of components. The charge pump output in this case is a series of positive and negative pulses, alternating in polarity.

$\tau > 0$: The charge pump will be on for a period of τ for every reference period, $1/fr$. Thus the average output of the charge pump would be: $\tau \bullet fr \bullet K\phi$. But this delay period, τ, can be associated with a phase delay by multiplying by 2π. So it can be seen that the time averaged output of the PFD is proportional to the phase error. Note that for two signals of the same frequency, their phase difference can always be expressed as a number between 0 and 2π. Therefore, the difference, τ, should always be less than $1/fn$ in this case.

Calculation of the Phase Detector Gain

To calculate the phase detector gain, it is necessary to consider the two extreme cases. When the phase error is $+2\pi$, it sources $K\phi$ current and when the phase error is -2π, it sinks $K\phi$ current. Within this range, the curve is linear. This means that the proper phase detector gain is $K\phi/2\pi$ (mA/rad). Although it is technically correct to divide by this factor of 2π, it is omitted in this book because it is multiplied by another of 2π which is used to convert the VCO gain from MHz/volt to Mrad/volt.

Analysis of The PFD for Two signals Differing in Frequency and Phase

Although this analysis of the PFD for a phase error is sufficient for most situations, some may be interested in how the phase detector behaves for two signals differing in frequency. This is of particular interest in the construction of lock detect circuits. For the purposes of this analysis, the following terms will be defined:

fr	The frequency of the signal coming from the crystal reference and then divided by R
ϕr	The phase of the fr signal at any given time
α	The initial phase of the fr signal
fn	The frequency of the signal coming from the VCO and then divided by N
ϕn	The phase of the fn signal at any given time
β	The initial phase of the fn signal
t	Elapsed time

Since frequency is the rate of change of the phase, it can be shown that:

$$\phi r = \alpha + fr \bullet t \qquad (3.1)$$

$$\phi n = \beta + fn \bullet t \qquad (3.2)$$

Looking in this perspective, the phase difference is obvious, therefore the time-averaged output of the phase detector for any given time, t, would be:

$$K\phi \bullet [\alpha - \beta + (fr - fn) \bullet t] \qquad (3.3)$$

The choice of t depends on whether or not $fr>fn$ or $fr<fn$. Without loss of generality, it will be assumed that $fr>fn$, if it is the other case, then a similar reasoning can be used. If one considers the average current output over P periods, this is shown below.

$$\begin{cases} \dfrac{K\phi}{P} \bullet \left(\alpha - \beta + (fr - fn) \bullet \dfrac{P}{fr} \right) & fr > fn \\ \dfrac{K\phi}{P} \bullet \left(\alpha - \beta + (fn - fr) \bullet \dfrac{P}{fn} \right) & fr < fn \end{cases} \qquad (3.4)$$

Taking the limit as P approaches infinity gives the time-averaged output of the phase detector:

$$\begin{cases} K\phi \bullet \left(1 - \dfrac{fn}{fr} \right) & fr > fn \\ K\phi \bullet \left(1 - \dfrac{fr}{fn} \right) & fr < fn \end{cases} \qquad (3.5)$$

When fr is an integer multiple of fn, these results in (3.5) above have been verified by computer simulation. However, for smaller frequency errors, it has been verified that the charge pump output is a function of the ratio of fr to fn, and that this increases linearly with the frequency error for small frequency errors only. In a real situation, the PLL is tracking the phase error, which causes some of these simulations to be somewhat unrealistic. The equations above serve as a rough guess at the duty cycle of the phase detector for a given frequency error. In a closed loop system, the PLL is tracking the phase error, and this can cause these estimates to be a little different than theoretically predicted.

The Continuous Time Approximation

Technically, the phase/frequency detector puts out a pulse width modulated signal and not a continuous current. However, it greatly simplifies calculations to approximate the charge pump current as a continuous current with a magnitude equal to the time-averaged value of these currents from the charge pump. This approximation is referred to as the continuous time approximation. This approximation loses accuracy as the comparison frequency approaches the loop bandwidth of the system. Despite this fact, this approximation holds very well in most cases and is used in order to derive the transfer functions that are necessary to analyze the PLL system. The discrete sampling effects that are not accounted for in the continuous time approximation introduce minor errors in the calculation of many performance criteria, such as the spurs, phase noise, and the transient response. These performance criteria will be discussed in greater detail in chapters to come, but the impact of these discrete sampling effects will be discussed here.

Discrete Sampling Effects on Spurs and Phase Noise

The impact of discrete sampling effects on spurs is typically not that great. However, if the loop bandwidth is wide relative to the comparison frequency, then sometimes a cusping effect can be seen. The discrete sampling action of the phase detector seems to have a much greater impact on phase noise. The phase detector/charge pump tends to be the dominant noise source in the PLL and it is these discrete sampling effects that cause the PFD to be nosier at higher comparison frequencies. Since a PFD with a higher comparison frequency has more corrections, it also puts out more noise, and this noise is proportional to the number of corrections. It is for this reason that the PFD noise increases as *10•log(Fcomp)*.

Discrete Sampling Effects on Loop Stability and Transient Response

The continuous time approximation holds when the loop bandwidth is small relative to the comparison frequency. If it is not, then theoretical predictions and actual results begin to differ and the PLL can even become unstable. Choosing the loop bandwidth to be $1/10^{th}$ of the comparison frequency is enough to keep one out of trouble, and when the loop bandwidth approaches around $1/3^{rd}$ the comparison frequency, simulation results show that this causes instability and the PLL to lose lock. In general, these effects should not be that much of a consideration.

The Phase/Frequency Detector Dead Zone

When the phase error is very small, there are problems with the phase/frequency detector responding to it correctly. Because the phase detector is made with real-world components, these gates have delays associated with them. When the time that the PFD would theoretically be on approaches the time delay of these components, then the output of the charge pump gets some added noise. This area of operation where the phase error is on the order of the component delays in the phase detector is referred to as the dead zone. Many PLLs have dead zone elimination circuitry ensures that the charge pump always comes on for some amount of time to avoid operating in the dead zone.

Conclusion

This chapter has discussed the PFD (Phase Frequency Detector) and has given some characterization on how it performs for both frequency and phase errors. For the phase error, it can be seen that the output is proportional to the phase error. For frequency errors, it can be seen that there is some output that is positively correlated with the frequency error.

The PFD is named so because it can detect differences in both phase and frequency. It also bypasses many limitations that are part of using a mixer or XOR phase detector, such as pull-in range, hold-in range, and steady state phase error.

References

Best, Roland E., 1995   *Phase-Lock Loop Theory, Design, Applications*, 3rd. ed, McGraw-Hill

Gardner, F.M.   *Phaselock Techniques*, 2nd ed., John Wiley & Sons, 1980

Gardner, F.M.,   *Charge-Pump Phase-Lock Loops*, IEEE Trans. Commun. vol. COM-28, pp. 1849 – 1858, Nov 1980

Chapter 4 Basic Prescaler Operation

Introduction

Until now, the *N* counter has been treated as some sort of black box that divides the VCO frequency and phase by *N*. If the output frequency of the VCO is low enough (on the order of 200 MHz or less), it can be implemented with a digital counter fabricated with a low frequency process, such as CMOS. It is desirable to implement as much of the *N* counter in CMOS as possible, for lower cost and current consumption. However, if the VCO frequency is much higher than this, then a pure CMOS counter is likely to have difficulty dealing with the higher frequency. To resolve this dilemma, prescalers are often used to divide down the VCO frequency to something that can be handled with lower the frequency processes. Prescalers often divide by some power of two, since this makes them easier to implement. The most common implementations of prescalers are single modulus, dual modulus, and quadruple modulus. Of these, the dual modulus prescaler is most commonly used.

Single Modulus Prescaler

For this approach, a single high frequency divider placed in front of a counter. In this case, $N = a \bullet P$, where *a* can be changed and *P* is fixed. One disadvantage of this prescaler is that only *N* values that are an integer multiple of *P* can be synthesized. Although the channel spacing can be reduced to compensate for this, doing so increases phase noise substantially. This approach also is popular in high frequency designs (>3 GHz) in which a fully integrated PLL cannot be fabricated totally in silicon. In this case, divide by two prescalers made with the GaAs or SiGe process can be used in conjunction with a PLL. Also, single modulus prescalers are sometimes used in older PLLs and low cost PLLs.

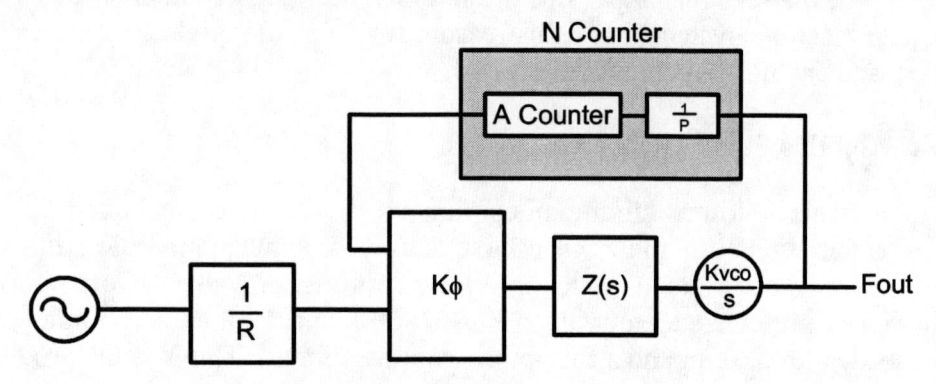

Figure 4.1 *Single Modulus Prescaler*

Dual Modulus Prescaler

In order not to sacrifice frequency resolution, a dual modulus prescaler is often used. These come in the form *P/(P+1)*. For instance, a *32/33* prescaler has *P = 32*. At first a fixed prescaler of size *P+1*, which is actually a prescaler of size *P* with a pulse swallow circuit, is engaged for a total of *a* cycles. Since the A counter activates the pulse swallow circuitry, it is often referred to as the swallow counter. It takes a total of **a•(P+1)** cycles for the A counter to count down to zero. Then the B counter starts counting down. Since it started with **b** counts, the remaining counts would be **(b – a)**. The size P prescaler is then switched in. This takes **(b-a)•P** counts to finish up the count, at which time, all of the counters are reset, and the process is repeated. From this the fundamental equations can be derived:

$$N = (P+1) \bullet a + P \bullet (b-a) = P \bullet b + a \quad (4.1)$$

$$b = N \ div \ P \ (\text{N divided by P, disregarding the remainder}) \quad (4.2)$$

$$a = N \ mod \ P \ (\text{The remainder when N is divided by P}) \quad (4.3)$$

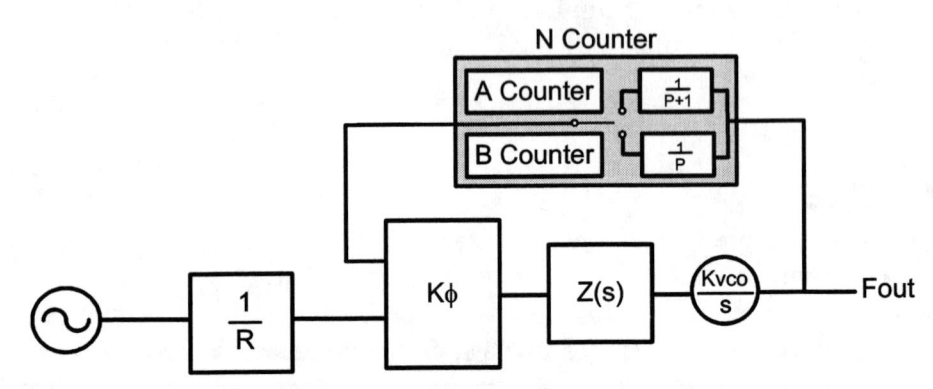

Figure 4.2 *Dual Modulus Prescaler*

Notice that **b>=a**, in order for proper operation, otherwise the B counter would prematurely reach zero and reset the system. For this reason, N values that yield **b<a** are called illegal divide ratios, and can not be achieved.

Quadruple Modulus Prescalers

In order to achieve a lower minimum continuous divide ratio, the quadruple modulus prescaler is often used. In the case of a quadruple modulus prescaler, there are four prescalers, but only three are used to produce any given N value. Commonly, but not always, these four prescalers are of values **P, P+1, P+4,** and **P+5**, and are implemented with a single pulse swallow circuit and a four-pulse swallow circuit. The N value produced is:

$$N = P \bullet c + 4 \bullet b + a \quad (4.4)$$
$$a = N \ mod \ P$$
$$c = N \ div \ P$$
$$b = \frac{N - c \bullet P - a}{4}$$

The following table shows the three steps and how the prescalers are used in conjunction to produce the required *N* value. Regardless of whether or not *b>=a*, the resulting *N* value is the same. Note that the *b>=a* restriction applies to the dual modulus prescaler, but not the quadruple modulus prescaler. The restriction for the quadruple modulus prescaler is *c >= max{a, b}*. *N* values that violate this rule are called illegal divide ratios.

Step	If b>=a		If b<a	
	Description	**Counts Required**	**Description**	**Counts Required**
1	The P+5 prescaler is engaged in order to decrement the A counter until a=0.	a•(P+5)	The P+5 prescaler is engaged in order to decrement the B counter until b=0.	b•(P+5)
2	The P+4 prescaler is engaged in order to decrement the B counter until b=0.	(b-a)•(P+4)	The P+1 prescaler is engaged in order to decrement the A counter until a=0.	(a-b)•(P+1)
3	The P prescaler is engaged in order to decrement the C counter until c=0.	(c-b)•P	The P prescaler is engaged in order to decrement the C counter until c=0.	(c-a)•P
	Total Counts	P•c+4•b+a	Total Counts	P•c+4•b+a

Table 4.1 *Typical Operation of a Quadruple Modulus Prescaler*

Minimum Continuous Divide Ratio

It turns out that for the dual modulus and quadruple modulus prescaler that all N values that are above a particular value, called the minimum continuous divide ratio, will be legal divide ratios. For the dual modulus prescaler, this is easy to calculate. Because *a* is the result of taking a number modulus *P*, the maximum this number can be is *P-1*. It therefore follows that if *b>=P-1*, the N value is legal. Using similar logic, the minimum continuous divide ratio for the quadruple modulus prescaler can also be calculated.

P	Dual Modulus		Quadruple Modulus	
	Prescaler	**Minimum Divide**	**Prescaler**	**Minimum Divide**
8	8/9	56	8/9/12/13	24
16	16/17	240	16/17/20/21	48
32	32/33	992	32/33/36/37	224
64	64/65	4032	64/65/68/69	960
128	128/129	16256	128/129/132/133	3968
P	P/(P+1)	P•(P-1)	P/(P+1)/(P+4)/(P+5)	P•max{3,max{P/4-1}}

Table 4.2 *Minimum Continuous Divide Ratios*

Exceptions to Legal Divide Ratios

For both the dual modulus and quadruple modulus prescalers, legal divide ratios and the minimum continuous divide ratio have been discussed. However, there are PLLs which put additional requirements beyond what the mathematics of calculating legal divide ratios imply. For instance, many PLLs with a dual modulus prescaler have the additional requirement that *b>=3*. Some fractional PLLs with a dual modulus prescaler have the requirement that *b>=a+2*. For delta-sigma PLLs, the N counter value is varied between many values, so all the values that it varies between must be legal. This effectively raises the minimum continuous divide ratio by a fixed number of counts.

Conclusion

For PLLs that operate at higher frequencies, prescalers are necessary to overcome process limitations. The basic operation of the single, dual, and quadruple modulus prescaler has been presented. Prescalers combine with the A, B, and C counters in order to synthesize the desired N value. Because of this architecture, not all N values are possible there will be N values that are unachievable. These values that are unachievable are called illegal divide ratios. If one attempts to program a PLL to use an illegal divide ratio, then the usual result is that the PLL will lock to the wrong frequency. The advantage of using higher modulus prescalers is that a greater range of N values can be achieved, particularly the lower N values. Fractional PLLs achieve a fractional N value by alternating the N counter between two or more values. In this case, it is necessary for all of these N values used to be legal divide ratios. I addition to legal divide ratios, there can be additional requirements that a PLL can have on the N counter value.

Many PLLs allow the designer more than one choice of prescaler to use. In the case of an integer PLL, the prescaler used usually has no impact on the phase noise, reference spurs, or lock time. This is assuming that the N value is the same. For some fractional N PLLs the choice of prescaler may impact the phase noise and reference spurs, despite the fact that the N value is unchanged.

Chapter 5   Fundamentals of Fractional *N* PLLs

Introduction

One popular misconception regarding fractional N PLLs is that they require different design equations and simulation techniques than are used for integer *N* PLLs. This is not the case. However, since fractional N PLLs contain compensation circuitry for the fractional spurs, they may exhibit some behaviors that would not be expected from an integer PLL. In addition to this, the performance will also be different, due to the fact that the *N* value is different. This chapter discusses some of the theoretical and practical behaviors of fractional N PLLs.

Theoretical Explanation of Fractional N

Fractional *N* PLLs differ from integer *N* PLLs in that some fractional N values are permitted. In general, a modulo *FDEN* fractional N PLL allows *N* values in the form of:

$$N = N_{int} + \frac{FNUM}{FDEN} \qquad (5.1)$$

Because the *N* value can now be a fraction, the comparison frequency can now be increased by a factor of *FDEN*, while still retaining the same channel spacing. Other than the architecture of the PLL, there could be other factors, such as illegal divide ratio, maximum phase detector limits, or the crystal frequency, that put limitations on how large *FDEN* can be. Illegal divide ratios can become a barrier to using a fractional N PLL, because reducing the N_{int} value may cause it to be an illegal divide ratio. Decreasing the N_{int} value corresponds to increasing the phase detector rate, which still must not exceed the maximum value in the datasheet specification. The crystal can also limit the use of fractional N, since the *R* value must be an integer. This implies that the crystal frequency must be a multiple of the comparison frequency.

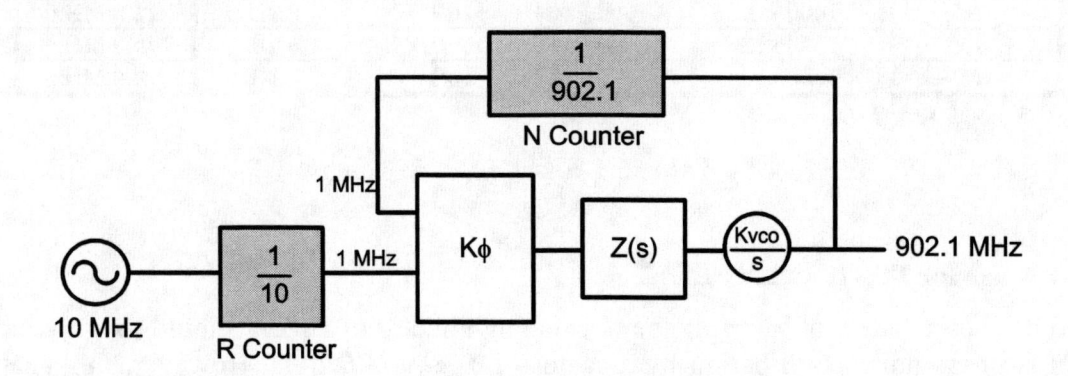

Figure 5.1 *Fractional N PLL Example Fractional N PLL Example*

Figure 5.1 shows an example of a fractional N PLL generating 902.1 MHz with **FDEN=10**. This PLL has a channel spacing of 100 kHz, but a reference frequency of 1 MHz. Now assume that the PLL tunes from 902 MHz to 928 MHz with a channel spacing of 100 kHz. The *N* value therefore ranges from 902.0 – 928.0. If a 32/33 dual modulus prescaler and the crystal frequency of 10 MHz were used, the *R* counter value would be an integer and all *N* values would be legal divide ratios. In this case, the crystal frequency and prescaler did restrict the use of fractional N. Now assume that this PLL of fractional modulus of **FDEN** is to be used and the PLL phase detector works up to 10 MHz. Below is a table showing if and how a modulo **FDEN** PLL could be used for this application. Since the comparison frequency is never bigger than 1600 MHz, there is no problem with the 10 MHz phase detector frequency limitation. In cases where the prescaler will not work, suggested values are given that will work. Since the quadruple modulus prescaler is able to achieve lower minimum continuous divide ratios, they tend to be more common in fractional N PLLs than integer *N* PLLs.

Fractional Modulo	Comparison Frequency	32/33 Prescaler Check	Prescaler Suggestion	10 MHz Crystal Check	Crystal Suggestion
1	100 kHz	OK		OK	
2	200 kHz	OK		OK	
3	300 kHz	OK		FAIL	14.4 MHz
4	400 kHz	OK		OK	
5	500 kHz	OK		OK	
6	600 kHz	OK		FAIL	6.0 MHz
7	700 kHz	OK		FAIL	7.0 MHz
8	800 kHz	OK		OK	
9	900 kHz	OK		FAIL	14.4 MHz
10	1000 kHz	FAIL	16/17	OK	
11	1100 kHz	FAIL	16/17	FAIL	11.0 MHz
12	1200 kHz	FAIL	16/17	FAIL	14.4 MHz
13	1300 kHz	FAIL	16/17	FAIL	13.0 MHz
14	1400 kHz	FAIL	16/17	FAIL	14.0 MHz
15	1500 kHz	FAIL	16/17	FAIL	15.0 MHz
16	1600 kHz	FAIL	16/17	FAIL	14.4 MHz

Table 5.1 *Fractional N Example*

Phase Noise for Fractional *N* PLLs

It will be shown later that lowering the N value by a factor of **FDEN** should roughly reduce the PLL phase noise contribution by a factor of **10•log(FDEN)**. However, this analysis disregards the fact that the fractional compensation circuitry can add significant phase noise. A good example is the National Semiconductor LMX2350. Theoretically, using this part in modulo 16 mode, one would expect a theoretical improvement of 12 db over its integer *N* counterpart, the LMX2330. At 3 V, the improvement is closer to 1 db. This is because the fractional circuitry adds about 11 db of noise. Using this part in modulo 8 mode at 3 V would actually yield a degradation of 2 db. At 4 V and higher operation, the fractional circuitry only adds 7 db, making this part more worthwhile. Depending on the method of fractional compensation used and the PLL, the added noise due to the fractional circuitry can be different. Many fractional N PLLs also have selectable prescalers, which can have a large impact on phase noise. For an integer part, choosing a different prescaler has no impact on phase noise. Also some parts allow the fractional compensation circuitry to be bypassed, which results in a fair improvement in phase noise at the expense of a large increase in the fractional spurs. For some applications, the loop bandwidth may be narrow enough to tolerate the increased fractional spurs.

Fractional Spurs for Fractional *N* PLLs

Since the reference spurs for a fractional N PLL are ***FDEN*** times the frequency offset away, they are often not a problem, since the loop filter can filter them more. However, fractional N PLLs also have fractional spurs, which are caused by imperfections in the fractional compensation circuitry. The first fractional spur is typically the most troublesome and occurs at ***1/FDEN*** times the comparison frequency, which is the same offset that the main reference spur occurs for the integer *N* PLL. As with phase noise, the fractional spur level is also dependent on the choice of prescaler and voltage. Recall from the reference spur chapter that the ***BasePulseSpur*** for the LMX2350 contains an added term, which depends on the output frequency. If the fractional numerator is set to one, then all the fractional spurs will be present. However, the kth fractional spur will be worst when the fractional numerator is equal to *k*. It is not necessarily true that switching from an integer PLL to a fractional PLL will result in reduced spur levels. Fractional *N* PLLs have the greatest chance for spur levels when the comparison frequency is low and the spurs in the integer PLL are leakage dominated. Fractional spurs are highly resistant to leakage currents. To confirm this, leakage currents up to 5 µA were induced to a PLL with 25 kHz fractional spurs (*FDEN=16, Fcomp=400 kHz*) and there was no observed degradation in spur levels.

Lock Time for Fractional *N* PLLs

There are two indirect ways that a fractional N PLL can yield improvements in lock time. The first situation is where the fractional N part has lower spurs, thus allowing an increase in loop bandwidth. If the loop bandwidth is increased, then the lock time can be reduced in this way. The second, and more common, situation occurs when the discrete sampling rate of the phase detector is limiting the loop bandwidth. Recall that the loop bandwidth cannot be practically made much wider than $1/5^{th}$ of the comparison frequency. If the comparison frequency is increased by a factor of ***FDEN***, then the loop bandwidth can be increased. This is assuming that the spur levels are low enough to tolerate this increase in loop bandwidth.

Fractional *N* Architectures

The way that fractional N values are typically achieved is by toggling the *N* counter value between two or more values, such that the average *N* value is the desired fractional value. For instance, to achieve a fractional value of 100 1/3, the *N* counter can be made 100, then

100 again, then 101. The cycle repeats. The simplest way to do the fractional N averaging is to toggle between two values, but it is possible to toggle between three or more values. If more than two values are used then this is a delta sigma PLL architecture, which is discussed in the next chapter.

An accumulator is used to keep track of the instantaneous phase error, so that the proper N value can be used and the instantaneous phase error can be compensated for (Best 1995). Although the average N value is correct, the instantaneous value is not correct, and this causes high fractional spurs. In order to deal with the spur levels, a current can be injected into the loop filter to cancel these. The disadvantage of this current compensation technique is that it is difficult to get the correct timing and pulse width for this correction pulse, especially over temperature. Another approach is to introduce a phase delay at the phase detector. This approach yields more stable spurs over temperature, but sometimes adds phase noise. In some parts that use the phase delay compensation technique, it is possible to shut off the compensation circuitry in order to sacrifice reference spur level (typically 15 db) in order to improve the phase noise (typically 5 db). The nature of added phase noise and spurs for fractional parts is very part specific.

Table 5.2 shows how a fractional N PLL can be used to generate a 900.2 MHz signal from a 1 MHz comparison frequency, using the phase delay technique. This corresponds to an N value of 900.2. Note that a 900.2 MHz signal has a period of 1.111 pS, and a 1 MHz signal has a period of 1000 nS.

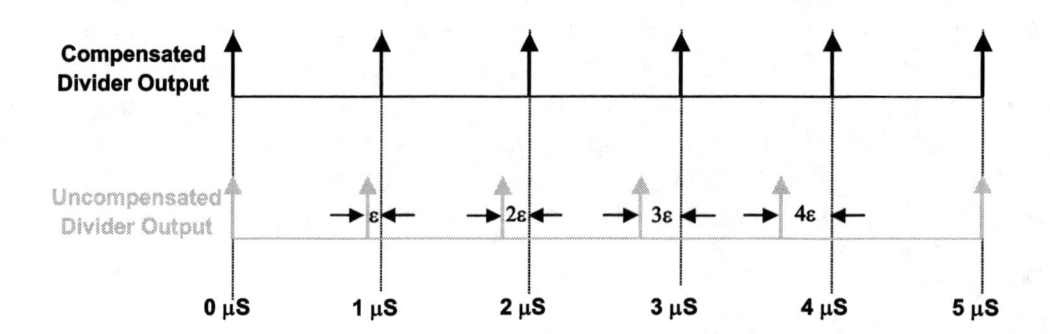

Figure 5.2 *Timing Diagram for Fractional Compensation*

Phase Detector Cycle	Accumulator (Cycles)	Overflow (Cycles)	Time for Rising Edge for Dividers (nS)		Phase Delay (nS)	Phase Delay (nS)
			Uncompensated	Compensated		
0	0.0	0	999.7778	0	0.222	0.222
1	0.2	0	1999.556	2000	0.444	0.444
2	0.4	0	2999.333	3000	0.667	0.667
3	0.5	0	3999.111	4000	0.889	0.889
4	0.8	0	4998.889	5000	1.111	1.111
5	0.0	1	999.7778	6000	0.222	0.222

Table 5.2 *Fractional N Phase Delay Compensation Example*

In Table 5.2, only the VCO cycles that produce a signal out of the N counter are accounted for. Note that the phase delay is calculated as follows:

$$Phase\ Delay = \frac{1}{VCO\ Frequency} \bullet (Accumulator\ Value + Overflow\ Value) \qquad (5.2)$$

When the accumulator value exceeds one, then an overflow count of one is produced, the accumulator value is decreased by one, and the next VCO cycle is swallowed (Best 1995). Note that in Table 5.2, this whole procedure repeats every 5 phase comparator cycles, which corresponds to 4501 VCO cycles.

Conclusion

The behavior and benefits of the fractional N PLL have been discussed. Although the same theory applies to a fractional *N* PLL as an integer PLL, the fractional N compensation circuitry can cause many quirky behaviors that are typically not seen in integer *N* PLLs. For instance, the National Semiconductor LMX2350 PLL has a dual modulus prescaler that requires *b>=a+2*, instead of *b>=a*, which is typical of integer *N* PLLs. Phase noise and spurs can also be impacted by the choice of prescaler as well as by the Vcc voltage to the part. Fractional *N* PLLs are not for all applications and each fractional N PLL has its own tricks to usage.

Reference

Best, Roland E., *Phase-Locked Loop Theory, Design, and Applications*, 3rd ed, McGraw-Hill, 1995

Chapter 6 Delta Sigma Fractional N PLLs

Introduction

Actually, the first order delta sigma PLL has already been discussed in the previous chapter. The traditional fractional PLL alternates the N counter value between two values in order to achieve a counter value that is something in between. However, the first order delta sigma PLL is often considered a trivial case, and people usually mean at least second order when they refer to delta sigma PLLs. For purposes of discussion, traditional fractional N PLL will refer to a PLL with first order delta sigma order, and a delta sigma PLL will be intended to refer to fractional PLLs with a delta sigma order of two or higher, unless otherwise stated.

Delta Sigma Modulator Order

Although it is theoretically possible for analog compensation schemes to completely eliminate the fractional spurs without any ill effects, there are many issues with using them in real world applications. Schemes involving current compensation tend to be difficult to optimize to account for variations in process, temperature, and voltage. Schemes involving a time delay tend to add phase noise. In either case, analog compensation has its disadvantages. Another drawback of traditional analog compensation is that architectures that use it tend to get much more complicated as the fractional modulus gets larger. Although it is possible to have a traditional PLL that uses no analog compensation, doing so typically sacrifices on the order of 20 dB spurious performance, although this is very dependent on which PLL is used.

Delta sigma PLLs have no analog compensation and reduce fractional spurs using digital techniques in order to try to bypass a lot of the issues with using traditional analog compensation. The delta sigma PLL reduces spurs by alternating the N counter between more than two values. The impact that this has on the frequency spectrum is that it pushes the fractional spurs to higher frequencies that can be filtered more by the loop filter.

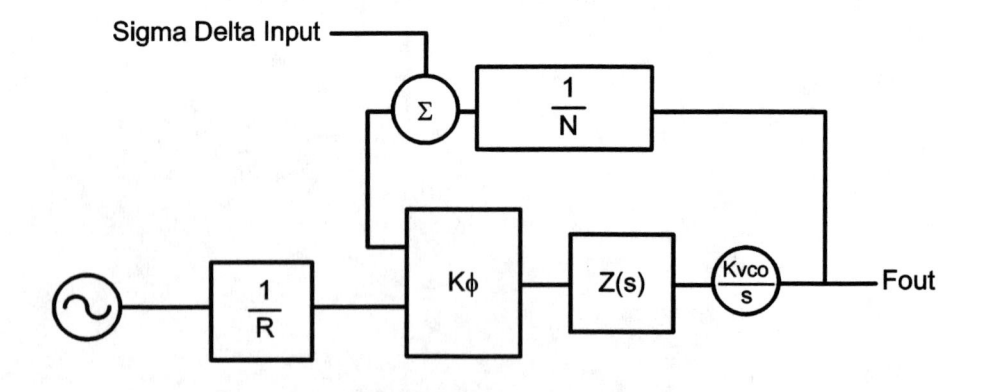

Figure 6.1 *Delta Sigma PLL Architecture*

Delta Sigma Order	Delta Sigma Input
1 (Traditional Fractional PLL without Compensation)	0, 1
2nd	-2, -1, 0, 1
3rd	-4, -3, -2, -1, 0, 1, 2, 3
4th	-8, ... +7
k^{th}	-2^k ... 2^k-1

Table 6.1 *Delta Sigma Modulator Example*

For example, consider a PLL with an N value of 100.25 and a comparison frequency of 1 MHz. A traditional fractional N PLL would achieve this by alternating the N counter values between 100 and 101. A 2nd order delta sigma PLL would achieve this by alternating the N counter values between 98, 99, 100, and 101. A 3rd order delta sigma PLL would achieve this by alternating the N counter values between 96, 97, 98, 99, 100, 101, 102, and 103. In all cases, the average N counter value would be 100.25. Note that all of the N counter values that the delta sigma PLL generates in its sequence must be legal divide ratios. The first fractional spur would be at 250 kHz, but the 3rd order delta sigma PLL would theoretically have lower spurs than the 2nd order delta sigma PLL. If there was no compensation used on the traditional fractional N PLL, this would theoretically have the worst spurs, but with compensation, this would depend on how good the analog compensation was.

Generation of the Delta Sigma Modulation Sequence

The sequence generated by the delta sigma modulator is dependent on the structure and the order of the modulator. For this case, the problem is modeled as having an ideal divider with some unwanted quantization noise. In this case, the quantization noise represents the instantaneous phase error of an uncompensated fractional divider. Figure 6.2 contains expressions involving the Z transform, which is the discrete equivalent of the Laplace transform. The expression in the forward loop representations a summation of the accumulator, and the z^{-1} in the feedback path represents a 1 clock cycle delay.

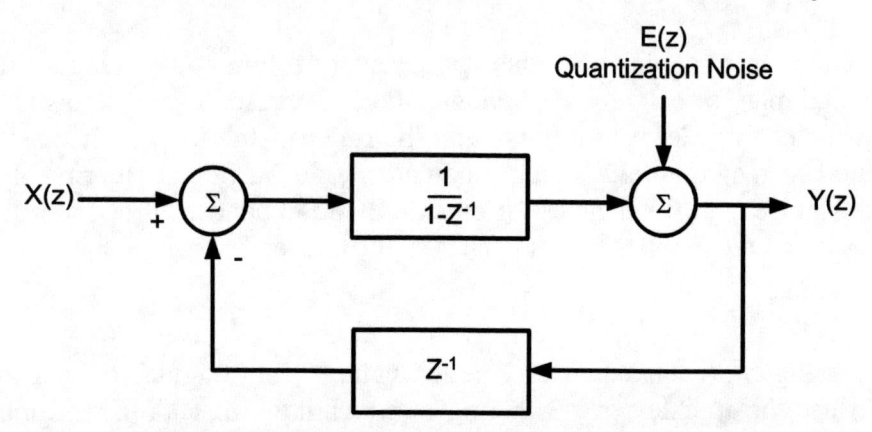

Figure 6.2 *The First Order Delta Sigma Modulator*

The transfer function for the above system is a follows:

$$Y(z) = X(z) + E(z) \cdot (1 - z^{-1}) \qquad (6.1)$$

Note that the error term transfer function means to take the present value and subtract away what the value was in the previous clock cycle. In other words, this is a form of digital high pass filtering. The following table shows what the values of this first order modulator would be for an *N* value of 900.2.

x[n]	Accumulator	e[n]	y[n]	N Value
0.2	0.2	-0.2	0	900
0.2	0.4	-0.4	0	900
0.2	0.6	-0.6	0	900
0.2	0.8	-0.8	0	900
0.2	1.0	-0.0	1	901
0.2	0.2	-0.2	0	900
0.2	0.4	-0.4	0	900
0.2	0.6	-0.6	0	900
0.2	0.8	-0.8	0	900
0.2	1.0	-0.0	1	901

Table 6.2 *Values for a First Order Modulator for N=900.2*

In general, the first order delta sigma modulator is considered a trivial case and delta sigma PLLs are usually meant to mean higher than first order. Although there are differences in the architectures, the general form of the transfer function for an nth order delta sigma modulator is:

$$Y(z) = X(z) + E(z) \cdot (1 - z^{-1})^n \qquad (6.2)$$

So in theory, higher order modulators push out the quantization noise to higher frequencies, that can be filtered more effectively by the loop filter. Because this noise pushed out grows at higher frequencies as order n, it follows that the order of the loop filter needs to be one greater than the order of the delta sigma modulator. If insufficient filtering is used, then even though this noise is at frequencies far outside the loop bandwidth it can mix and make spurious products that are at much closer offsets to the carrier.

Dithering
In addition to using more than two *N* counter values, delta sigma PLLs may also use dithering to reduce the spur levels. Dithering is a technique of adding randomness to the sequence. For example, an *N* counter value of 99.5 can be achieved with the following sequence:

98, 99, 100, 101, ... (pattern repeats)

Note that this sequence is periodic, which may lead to higher fractional spurs. Another sequence that could be used is:

99, 100, 98, 101, 98, 99, 100, 101, 98, 101, 99, 100 ... (pattern repeats)

Both sequences achieve an average N value of 99.5, but the second one has less periodicity, which theoretically implies that more of the lower frequency fractional spur energy is pushed to higher frequencies.

The impact of dithering is different for every application, but its impact on the main fractional spurs tends to be minimal. However, delta-sigma PLLs can sub-fractional spurs that occur at a fraction of the channel spacing, that can be significantly impacted by dithering. In some cases, it can improve sub-fractional spur levels, while in other cases, it can make these spurs worse. One example where dithering can degrade spur performance is in the case where the fractional numerator is zero.

Conclusion

The delta sigma architecture can be used in fractional PLLs to reduce the fractional spurs. This architecture is based on modulating the N counter value to reduce the fractional spurs. Many of the problems that are inherent to analog compensation used in the traditional fractional N PLL, such as added phase noise and spurs that vary over process and temperature, are overcome. In the process of reducing the main fractional spurs, sub-fractional spurs are sometimes introduced, depending on the application.

Reference

Connexant Application Note *Delta Sigma Fractional N Synthesizers Enable Low Cost, Low Power, Frequency Agile Software Radio*

Chapter 7 The PLL as Viewed from a System Level

Introduction

This chapter discusses, on a very rudimentary level, how a PLL could be used in a typical wireless application. It also briefly discusses the impact of phase noise, reference spurs, and lock time on system level performance.

Typical Wireless Receiver Application

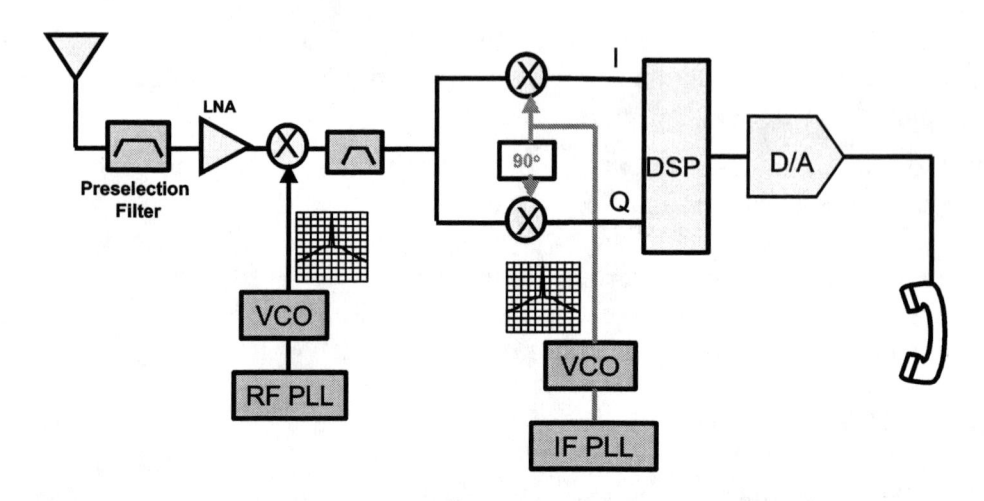

Figure 7.1 *Typical PLL Receiver Application*

General Receiver Description

In the above diagram, there are several different channels being received at the antenna, each one with a unique frequency. The first PLL in the receiver chain is tuned so that the output from the mixer is a constant frequency. The signal is then easier to filter and deal with since it is a fixed frequency from this point onwards, and because it is also lower in frequency. The second PLL is used to strip the information from the signal. Other than the obvious parameters of a PLL such as cost, size, and current consumption, there are three other parameters that are application specific. These parameters are phase noise, reference spurs, and lock time and are greatly influenced by the loop filter components. For this reason, these performance parameters are not typically specified in a datasheet, unless the exact application, components, and design parameters are known.

Phase Noise, Reference Spurs, and Lock Time as They Relate to This System

Phase noise refers to noise generated by the PLL. It can increase the bit error rates and degrade the signal to noise ratio of the system. This is discussed in depth in later chapters. Also, phase noise can mix with signals in order to create undesired noise products. Spurs are unwanted noise sidebands that can occur at multiples of the channel spacing, and can be translated by a mixer to the desired signal frequency. They can mask or degrade the desired signal. Lock time is the time that it takes for the PLL to change frequencies. It is dependent on the size of the frequency change and what frequency error is considered acceptable. When the PLL is switching frequencies, no data can be transmitted, so lock time of the PLL must lock fast enough as to not slow the data rate. Lock time can also be related to power consumption. For some systems, the PLL does not need to be powered up all the time, but only when data is transmitted or received. During other times, the PLL and many other RF components can be off. If the PLL lock time is less, then that allows systems like this to spend more time with the PLL powered down and therefore current consumption is reduced. Phase noise, reference spurs, and lock time are discussed in great depth in the rest of this book.

Analysis of Receiver System

For the receiver shown in Figure 7.1 , the PLL that is closest to the antenna is typically the most challenging from a design perspective, due to the fact that it is higher frequency and is tunable. Since this PLL is tunable, there is typically a more difficult lock time requirement, which in turn makes it more challenging to meet spur requirements as well. In addition to this, the requirements on this PLL are also typically stricter because the undesired channels are not yet filtered out from the antenna.

The second PLL has less stringent requirements, because it is lower frequency and also it is often fixed frequency. This makes lock time requirements easier to meet. There is also a trade off between lower spur levels and faster lock times for any PLL. So if the lock time requirements are relaxed, then the reference spur requirements are also easier to meet. Note also that since the signal path coming to the second PLL has already been filtered, the lock time and spur requirements are often less difficult to meet.

Example of an Ideal System with an Ideal PLL

For this example, assume all the system components are ideal. All mixers, LNAs and filters have 0 dB gain and noise figure. All filters are assumed to have an idea "brick wall" response. The PLL is assumed to put out a pure signal and have zero lock time.

Receive Frequency	869.03 – 893.96 MHz
RF PLL Frequency	783.03 – 807.96 MHz
IF PLL Frequency	86 MHz
Channel Spacing	30 kHz
Number of Channels	831
IF PLL Frequency	240 MHz

Table 7.1 *RF System Parameters*

The received channel will be one of the 831 channels. The channels will be designated 0 to 830, where channel 0 is at 869.03 MHz and channel 830 is at 893.96 MHz. Suppose the frequency to be received is channel 453 at 888.62 MHz. This frequency comes in through the antenna, filter, and LNA and is presented to the first mixer. The RF PLL frequency is then programmed to 802.62 MHz. The output of the mixer is therefore the sum and difference of these two frequencies, which would be 1691.24 MHz and 86 MHz. The filter afterwards filters out the high frequency signal so that only the 86 MHz signal passes through. This 86 MHz signal is then down converted to baseband with the IF PLL frequency, which is a fixed 86 MHz.

Ideal System with a Non-Ideal PLL

Now assume the same system as before, but now the RF and IF PLL puts out phase noise and spurs. Assume that the RF PLL takes 1 mS to change frequencies and the IF PLL takes 10 mS to change channels. For this application, the fact that the IF PLL takes 10 mS to change channels really does not have any impact on system performance. What this means is that once the phone is turned on, it takes an extra 10 mS to power up. Because the IF PLL never changes frequency, this is the only time this lock time comes into play.

Now the 1 mS lock time on the RF PLL has a greater impact. If a person was using their cell phone and it was necessary to change the channel, then this lock time would matter. This might happen if the user was leaving a cell and entering another cell and the channel they were on was in use. Also, sometimes there is a supervisory channel that the cell phone needs to periodically switch to in order to receive and transmit information to the network. This is the factor that drives the lock time requirement for the PLL in the IS-54 standard, after which this example was modeled. The time needed to switch back and forth to do this needs to be transparent to the user and no data can be transmitted or received when the PLL is switching frequencies.

In the case of spurs, they will be at 30 kHz offset from the carrier. This would be at frequencies of 802.59 MHz and 802.65 MHz. Now the strength of these signals would be much less than that at 802.62 MHz, but still they would be there. Now if there were any other users on the system, these spurs could cause problems. For instance, a user at 888.59 MHz or 888.65 MHz would mix with these spurs to also form an unwanted noise signal at 86 MHz. In actuality, there are spurs at every multiply of 30 kHz from the carrier, so there are more possibilities for noise signals at 86 MHz, but the ones mentioned above would be the worst-case.

Because the phase noise is a continuous function of the offset frequency, it can mix in many more ways to produce jammer signals. Phase noise as well as spurs cause an increase in the RMS phase error as well. This will be discussed later. The next three figures are an example of what impact phase noise and spurs can have on system performance. Note that the phase noise of the RF PLL is translated onto the output signal of the mixer. The undesired channel at 888.65 MHz causes two unwanted signals. The first one is at 802.62 MHz, which degrades the signal to noise ratio. Another product is caused by this 802.62 MHz signal mixing with the main signal. However, this is outside of the information bandwidth of the signal and would be attenuated by the channel selection filter after the mixer.

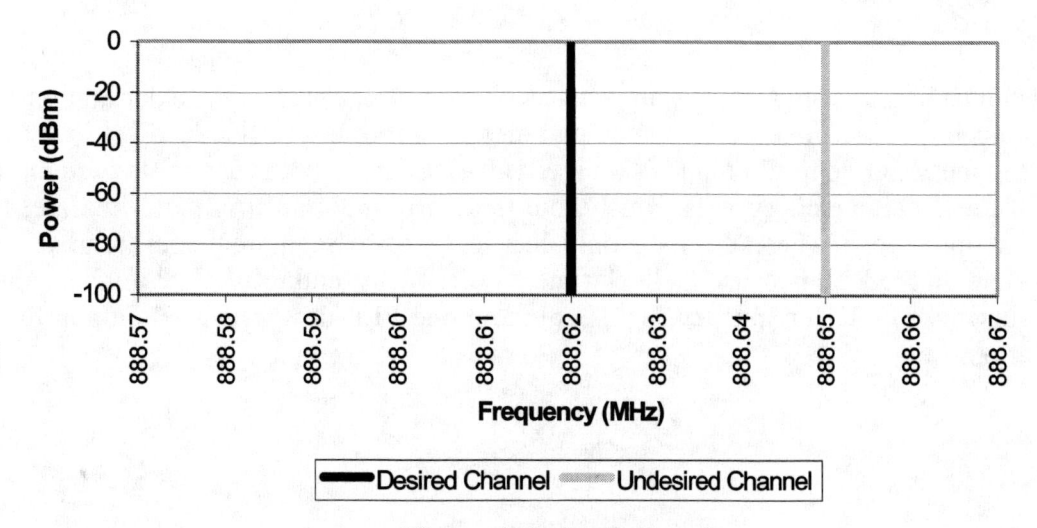

Figure 7.2 *Input Signal*

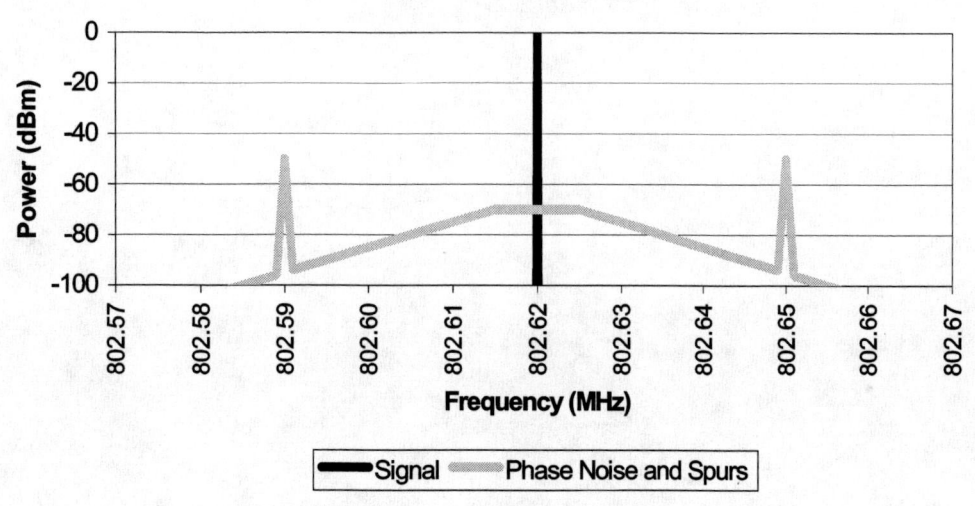

Figure 7.3 *Signal with Noise from RF PLL*

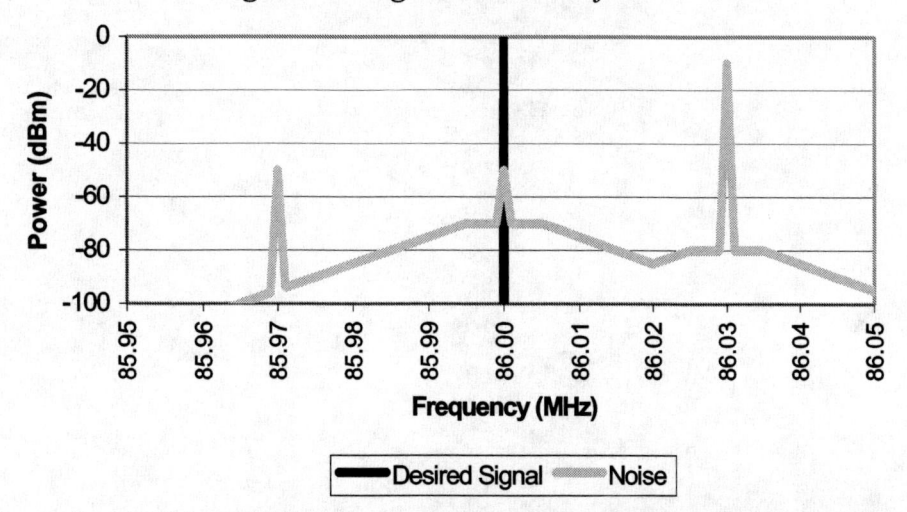

Figure 7.4 *Output Signal from Mixer*

Conclusion

This chapter has investigated the impacts of phase noise, spurs, and lock time on system performance. These three performance parameters are greatly influenced by many factors including the VCO, loop filter, and N divider value. Of course it is desirable to minimize all three of these parameters simultaneously, but there are important trade-offs that need to be made. Applications where the PLL only has to tune to fixed frequency tend to be less demanding on the PLL because the lock time requirements tend to be very relaxed, allowing one to optimize more for spur levels. There is no one PLL design that is optimal for every application.

PLL Performance and Simulation

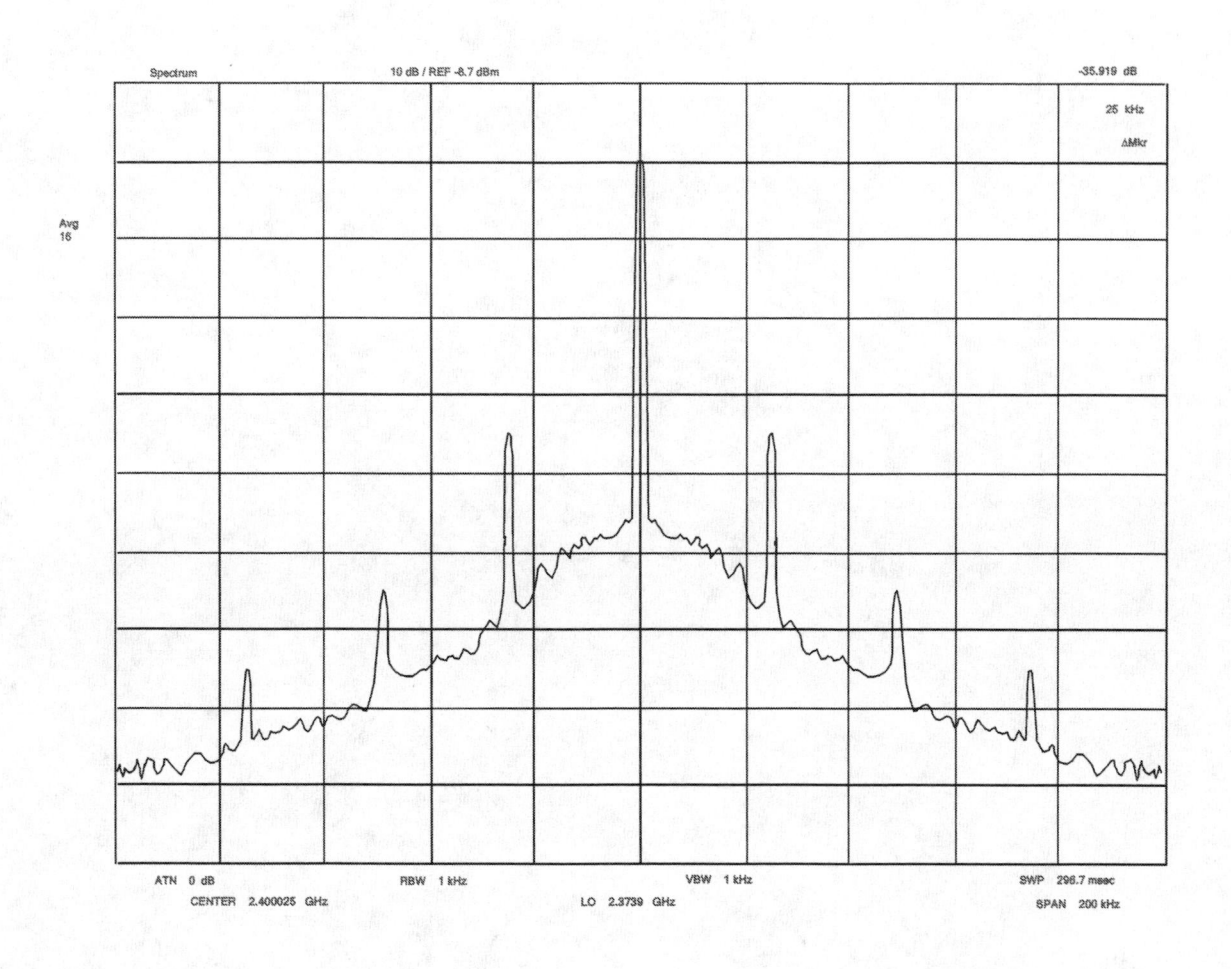

Chapter 8 Introduction to Loop Filter Coefficients

Introduction

This chapter introduces notation used to describe loop filter behavior throughout this book. The loop filter transfer function will be defined as the voltage at the tuning port of the VCO divided by the current at the charge pump that caused it. In the case of a second order loop filter, it is simply the impedance. The transfer function of any PLL loop filter can be described as follows:

$$Z(s) = \frac{1 + s \bullet T2}{s \bullet (A3 \bullet s^3 + A2 \bullet s^2 + A1 \bullet s + A0)} \quad (8.1)$$

$$T2 \quad = \quad R2 \bullet C2 \quad (8.2)$$

$A0$, $A1$, $A2$, and $A3$ are the filter coefficients of the filter. In the case of a second order loop filter, $A2$ and $A3$ are zero. In the case of a third order loop filter, $A3$ is zero. If the loop filter is passive, then $A0$ is the sum of the capacitor values in the loop filter. In this book, there are two basic topologies of loop filter that will be presented, passive and active. Although, there are multiple topologies presented for the active filter, only one is shown here, since this is the preferred approach.

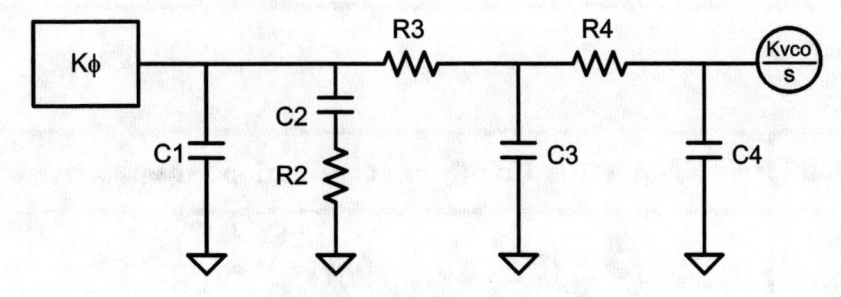

Figure 8.1 *Passive Loop Filter*

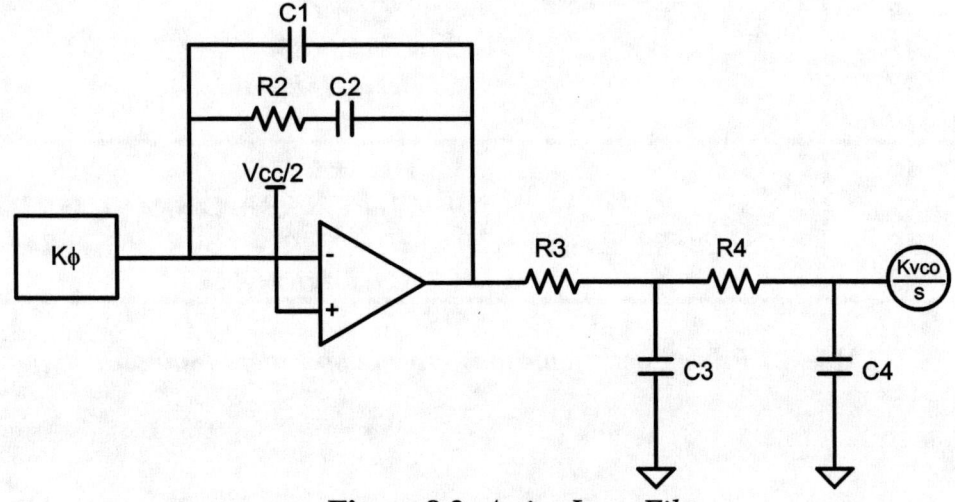

Figure 8.2 *Active Loop Filter*

PLL Performance, Simulation, and Design © 2003, Third Edition

Calculation of Filter Coefficients

Realize that although equations for the 2nd and 3rd order are shown, they can be easily derived from the 4th order equations by setting the unused component values to zero. In order to simplify calculations later on, the filter coefficients will be referred to many times, so it is important to be very familiar how to calculate them.

Filter Order	Symbol	Filter Coefficient Calculation for a Passive Filter
2	$A0$	$C1 + C2$
	$A1$	$C1 \cdot C2 \cdot R2$
	$A2$	0
	$A3$	0
3	$A0$	$C1 + C2 + C3$
	$A1$	$C2 \cdot R2 \cdot (C1+C3) + C3 \cdot R3 \cdot (C1+C2)$
	$A2$	$C1 \cdot C2 \cdot C3 \cdot R2 \cdot R3$
	$A3$	0
4	$A0$	$C1 + C2 + C3 + C4$
	$A1$	$C2 \cdot R2 \cdot (C1+C3+C4) + R3 \cdot (C1+C2) \cdot (C3+C4) + C4 \cdot R4 \cdot (C1+C2+C3)$
	$A2$	$C1 \cdot C2 \cdot R2 \cdot R3 \cdot (C3+C4) + C4 \cdot R4 \cdot (C2 \cdot C3 \cdot R3 + C1 \cdot C3 \cdot R3 + C1 \cdot C2 \cdot R2)$
	$A3$	$C1 \cdot C2 \cdot C3 \cdot C4 \cdot R2 \cdot R3 \cdot R4$

Table 8.1 *Filter Coefficients for Passive Loop Filters*

Filter Order	Symbol	Filter Coefficient Calculation for an Active Filter
2	$A0$	$C1 + C2$
	$A1$	$C1 \cdot C2 \cdot R2$
	$A2$	0
	$A3$	0
3	$A0$	$C1 + C2$
	$A1$	$C1 \cdot C2 \cdot R2 + C3 \cdot R3 \cdot (C1+C2)$
	$A2$	$C1 \cdot C2 \cdot C3 \cdot R2 \cdot R3$
	$A3$	0
4	$A0$	$C1 + C2$
	$A1$	$C1 \cdot C2 \cdot R2 + (C1+C2) \cdot (C3 \cdot R3 + C4 \cdot R4 + C4 \cdot R3)$
	$A2$	$C3 \cdot C4 \cdot R3 \cdot R4 \cdot (C1+C2) + C1 \cdot C2 \cdot R2 \cdot (C3 \cdot R3 + C4 \cdot R4 + C4 \cdot R3)$
	$A3$	$C1 \cdot C2 \cdot C3 \cdot C4 \cdot R2 \cdot R3 \cdot R4$

Table 8.2 *Filter Coefficients for Active Loop Filters (Standard Type)*

The calculation of the zero, *T2*, is the same for active and passive filters and independent of loop filter order:

$$T2 = C2 \bullet R2 \quad (8.3)$$

Calculation of Loop Filter Coefficients from Loop Filter Poles

In order to get a more intuitive feel of the loop filter transfer function, it is often popular to express this in terms of poles and zeroes. If one takes the reciprocal of the poles or zero values, then they get the corresponding frequency in radians. In the case of a fourth order passive loop filter, it is possible to get complex poles.

$$Z(s) = \frac{1 + s \bullet T2}{s \bullet A0 \bullet (1 + s \bullet T1) \bullet (1 + s \bullet T3) \bullet (1 + s \bullet T4)} \quad (8.4)$$

Once the loop filter time constants are known, it is easy to calculate the loop filter coefficients. The relationships between the time constants and filter coefficients are shown below.

$$\frac{A1}{A0} = T1 + T3 + T4 \quad (8.5)$$

$$\frac{A2}{A0} = T1 \bullet T3 + T1 \bullet T4 + T3 \bullet T4$$

$$\frac{A3}{A0} = T1 \bullet T3 \bullet T4$$

Calculation of Loop Filter Poles from Loop Filter Coefficients

In this section, a general approach for calculating the loop filter poles that works for all loop filter will be presented. Although this approach works for all loop filters, the calculations can be simplified in the case of an active loop filter. The method changes depending on whether the loop filter is second, third, or fourth order.

Second Order Loop Filter

The calculation of the pole, *T1* is trivial in this case.

$$T1 = \frac{A1}{A0} \quad (8.6)$$

Third Order Loop Filter

It is common to approximate the passive third order poles with the active third order poles. In order to solve exactly, it is necessary to solve a system of two equations and two unknowns.

$$T1+T3=\frac{A1}{A0} \tag{8.7}$$

$$T1 \cdot T3 = \frac{A2}{A0}$$

$$T1, T3 = \frac{A1 \pm \sqrt{A1^2 - 4 \cdot A0 \cdot A2}}{2 \cdot A0} \tag{8.8}$$

Fourth Order Loop Filter

For the passive fourth order loop filter, the time constants satisfy the following system of equations:

$$T1+T3+T4 = \frac{A1}{A0} \tag{8.9}$$

$$T1 \cdot T3 + T3 \cdot T4 + T1 \cdot T4 = \frac{A2}{A0}$$

$$T1 \cdot T3 \cdot T4 = \frac{A3}{A0}$$

If one uses the first equation to eliminate the variable *T1*, the result is as follows:

$$x^2 - \frac{A1}{A0} \cdot x + \frac{A2}{A0} = y \tag{8.10}$$

$$y \cdot \left(x - \frac{A1}{A0} \right) = \frac{A3}{A0}$$

where

$$x = T3+T4$$
$$y = T3 \cdot T4$$

Solving the first equation for y and substituting in the second equation yields:

$$x^3 - 2 \cdot \frac{A1}{A0} \cdot x^2 + \left(\frac{A1^2}{A0^2} + \frac{A2}{A0} \right) \cdot x + \left(\frac{A3}{A0} - \frac{A1 \cdot A2}{A0^2} \right) = 0 \tag{8.11}$$

Although a closed form solution to the third order cubic equation exists, there will always be at least one real root. It is easiest to find this root numerically. Once this is found, then y can be found, and the poles can be found in a similar way as in the third order passive filter. One rather odd artifact of the fourth order passive filter is that it is possible for the poles *T3* and *T4* to be complex and yet still have a real-world working loop filter. Although this can happen, it is not very common.

$$T3, T4 = \frac{x \pm \sqrt{x^2 - 4 \bullet y}}{2} \tag{8.12}$$

$$T1 = \frac{A3}{A0 \bullet y} \tag{8.13}$$

Simplification for Active Filters

Although the methods already presented work just fine for active filters, these equations can be simplified. The simplification comes from the fact that the pole, *T1*, can be easily calculated due to the isolation provided from the active device.

$$T1 = \frac{a1}{a0} \tag{8.14}$$

In this case, the filter coefficients are intentionally left lower case. These coefficients are calculated as they would be in the case of a second order filter. The components *C3*, *C4*, *R3*, and *R4* are all set to zero for the purposes of the calculation of the time constant *T1*.
In the case of a third order active filter, the calculation of the pole ,*T3*, can easily be found using equation the first equation in equation set (8.9). In the case of a fourth order active filter, one get a set of two equations and two unknowns for *T1* and *T3*.

$$x = \frac{A1}{A0} - T1 \tag{8.15}$$

$$y = \frac{A3}{A0 \bullet T1}$$

$$T3, T4 = \frac{x \pm \sqrt{x^2 - 4 \bullet y}}{2}$$

Conclusion

It is common to discuss a loop filter in terms of poles and zeros. However, it turns out that it greatly simplifies notation to introduce the filter coefficients as well. In addition to this, the filter coefficients are much easier to calculate for higher order filters. The zero, *T2*, is always calculated the same way, but the calculations for the poles depends on the loop filter order and whether or not the loop filter is active or passive. The purpose of this chapter was to make the reader familiar with the filter coefficients, *A0*, *A1*, *A2*, and *A3*, since they will be used extensively throughout this book.

Chapter 9   Introduction to PLL Transfer Functions and Notation

Introduction

This chapter introduces fundamental transfer functions and notation for the PLL that will be used throughout this book. A clear understanding of these transfer functions is critical in order to understand spurs, phase noise, lock time, and PLL design.

PLL Basic Structure

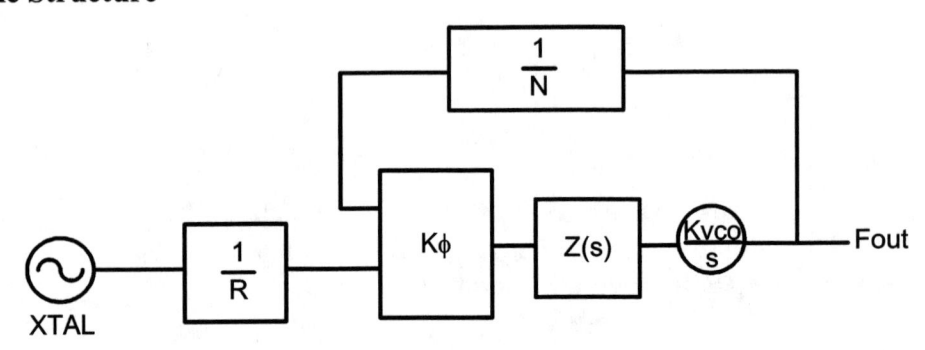

Figure 9.1 *Basic PLL Structure*

Introduction of Transfer Functions

The open loop transfer function is defined as the transfer function from the phase detector input to the output of the PLL. Note that the VCO gain is divided by a factor of s. This is to convert output frequency of the VCO into a phase. Technically, this transfer function is the phase of the PLL output divided by the phase presented to the phase detector, assuming the other input, ϕr, is a constant zero phase. The open loop transfer function is shown below:

$$G(s) = \frac{K\phi \cdot Kvco \cdot Z(s)}{s} \qquad (9.1)$$
$$s = 2\pi \cdot j \cdot f$$

The N counter value is the output frequency divided by the comparison frequency. Although the mathematics involved in defining H as the reciprocal of N may not be that impressive, it does make the equations look more consistent with notation used in classical control theory textbooks.

$$H = \frac{1}{N} \qquad (9.2)$$

The closed loop transfer function takes into account the whole system and does not assume that the phase of one of the phase detector inputs is fixed at a constant zero phase.

$$CL(s) = \frac{G(s)}{1+G(s) \bullet H} \qquad (9.3)$$

The transfer function in (9.3) involves an output phase divided by an input phase. In other words, it is a phase transfer function. However, the frequency transfer function would be exactly the same. If one is considering an input frequency, this could be converted to a phase by dividing by a factor of s, then it is converted to a phase. At the output, one would multiply by a factor of s to convert the output phase to a frequency. So both of these factors cancel out, which proves that the phase transfer functions and frequency transfer functions are the same. By considering the change in output frequency produced by introducing a test frequency at various points in the PLL loops, all of the transfer functions can be derived.

Source	Transfer Function
Crystal Reference	$\dfrac{1}{R} \bullet \dfrac{G(s)}{1+G(s) \bullet H}$
R Divider	$\dfrac{G(s)}{1+G(s) \bullet H}$
N Divider	$\dfrac{G(s)}{1+G(s) \bullet H}$
Phase Detector	$\dfrac{1}{K\phi} \bullet \dfrac{G(s)}{1+G(s) \bullet H}$
VCO	$\dfrac{1}{1+G(s) \bullet H}$

Table 9.1 *Transfer functions for various parts of the PLL*

Analysis of Transfer Functions

Note that the crystal reference transfer function has a factor of *1/R* and the phase detector transfer function has a factor of *1/Kϕ*. It is also true that the phase detector noise, *N* divider noise, *R* divider noise, and the crystal noise all contain a common factor in their transfer functions. This common factor is given below.

$$\frac{G(s)}{1+G(s) \bullet H} \qquad (9.4)$$

All of these noise sources will be referred to as in-band noise sources. The loop bandwidth, ωc, and phase margin, ϕ, are defined as follows:

$$\|G(j\bullet\omega c)\bullet H\| = 1 \qquad (9.5)$$

$$180 - \angle G(j\bullet\omega c)\bullet H = \phi \qquad (9.6)$$

The loop bandwidth relates to the closed loop bandwidth of the PLL system, and the phase margin relates to the stability. If the phase margin is too low, the PLL system may become unstable. Another parameter of interest that will be of more interest in the loop filter design chapters is the gamma optimization factor, which is defined as follows:

$$\gamma = \frac{T2}{\omega c^2 \bullet A0} \qquad (9.7)$$

Using these definitions, and equations (9.1) and (9.2), and the fact that $G(s)$ is monotonically decreasing in s yields the following transfer function:

$$\frac{G(s)}{1+G(s)\bullet H} \approx \begin{cases} N & \text{For } \omega \ll \omega c \\ G(s) & \text{For } \omega \gg \omega c \end{cases} \qquad (9.8)$$

The VCO has a different transfer function:

$$\left\|\frac{G(s)}{1+G(s)\bullet H}\right\| \qquad (9.9)$$

Note that this transfer function can be approximated by:

$$\frac{1}{1+G(s)\bullet H} \approx \begin{cases} \dfrac{N}{G(s)} & \text{For } \omega \ll \omega c \\ 1 & \text{For } \omega \gg \omega c \end{cases} \qquad (9.10)$$

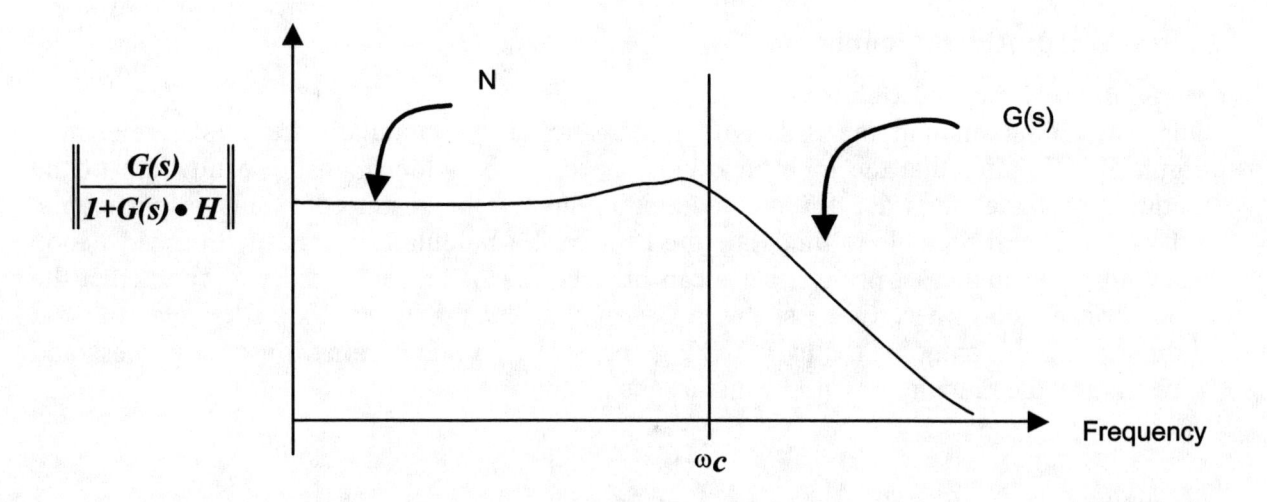

Figure 9.2 *Transfer Function Multiplying all Sources Except the VCO*

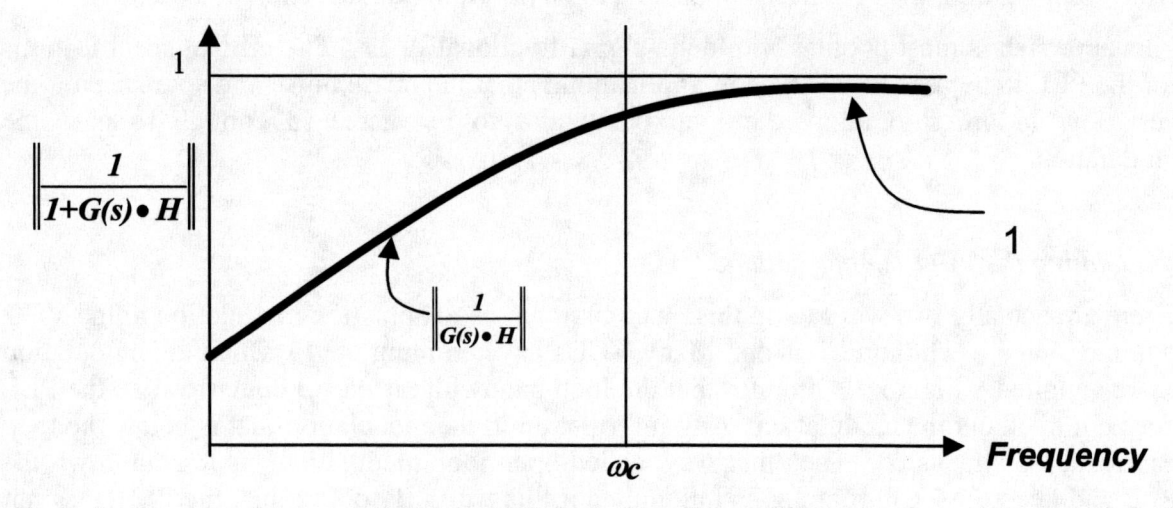

Figure 9.3 *Transfer Function for the VCO*

A Few Words About Modulation

Modulating the Crystal Reference

One way of modulating the PLL with information is to modulate the crystal reference. Figure 9.2 implies that the loop bandwidth needs to be wider than the bandwidth of the modulating signal in order for the modulation to not be distorted. This is preferable, although it is still possible to modulate the PLL with a modulating signal with a wider loop bandwidth. In this case, pre-emphasis can be used on the signal in order to counteract the roll-off of the loop filter. One issue with pre-emphasis is that this is loop filter specific, and if the loop filter component values, VCO gain, or charge pump gain vary over process and temperature, the modulation will be distorted slightly.

Modulating the N Counter Value

For this technique, the frequency of the VCO is changed by modulating the N counter value. In its crudest form, this is simply switching the VCO between two frequencies. Figure 9.2 implies that the bit rate needs to be less than the loop bandwidth, unless some sort of pre-emphasis is used. Although it technically is possible to use two values, this is typically not done because the spectral efficiency is low.

If the data is unshaped, then large spectral lobes will appear, and there can also be problems with intersymbol interference (ISI). If a fractional N PLL with a fine enough tuning resolution is used, then the data can be shaped by changing the N counter value. For this type of application, the PLL needs to be able to be programmed fast enough to do this.

This approach is most popular with delta-sigma fractional PLLs. The critical specifications for the PLL to be used in this type of application is that the phase noise and spurs need to be very low in order to allow the loop bandwidth to be increased enough to pass the modulation.

Modulating the VCO Tuning Voltage

There are actually two ways to do this. The first way is to apply the modulation to the VCO tuning voltage when the PLL is on. Many VCOs have a tuning pin to which the modulation can be applied. Figure 9.3 implies that the loop bandwidth needs to be narrow, so the PLL does not track out the modulation. Any information in the modulation that is below the loop bandwidth is attenuated. The other way, called open loop modulation, is to shut down the PLL and keep the VCO running and modulate it this way. By doing this, the PLL does not interfere with the modulated signal, but the frequency will eventually drift away from where it should be and then the PLL needs to be turned on again.

Dual Port Modulation

For dual port modulation, both the VCO and the crystal reference are modulated simultaneously. The advantage of this approach is that both the low frequencies and high frequencies can be passed. The disadvantage is the complexity and cost.

Scaling Properties of PLL Loop Filters

In order to discover these properties, it is easier to expand the expression for the closed loop transfer function and introduce a new variable, **K**.

$$\frac{G(s)}{1+G(s)/N} = \frac{K \cdot N \cdot (1+s \cdot T2)}{s^5 \cdot A3 + s^4 \cdot A2 + s^3 \cdot A1 + s^2 \cdot A0 + s \cdot K \cdot T2 + K} \qquad (9.11)$$

$$K = \frac{K\phi \cdot Kvco}{N}$$

Loop Gain Constant

The loop gain constant, **K**, is dependent on the charge pump gain, VCO gain, and **N** counter value. From the above equation, one can see that these three parameters can be changed without impacting the transfer functions, provided that the loop gain constant is held constant. For instance, if the charge pump gain and **N** counter value are doubled, yet the VCO gain is held constant, the loop filter transfer function remains unchanged.

Scaling Property of Components

Even though loop filter design has not been discussed yet, it is not premature to show how to scale loop filter components. The first step involves understanding how the loop filter coefficients change with the loop filter values. If one was to change all capacitors by a factor of *x* and all resistors by a factor of *y*, then the impact on the loop filter coefficients can be found.

Loop Filter Coefficient	Proportionality ($\propto$) to x and y
A0	x
A1	$\propto x \cdot (x \cdot y)$
A2	$\propto x \cdot (x \cdot y)^2$
A3	$\propto x \cdot (x \cdot y)^3$

Table 9.2 *Relationship of Loop Filter Coefficients to Component Scaling Factors*

Equation	Implication
$x \propto \dfrac{K}{Fc^2}$	The loop filter capacitors should be chosen proportional to the loop gain divided by the loop bandwidth squared.
$y \propto \dfrac{Fc}{K}$	The loop filter resistors should be chosen proportional to the loop bandwidth over the loop gain.

Table 9.3 *The Rule for Scaling Components*

Table 9.3 shows the fundamental rule for scaling components. There are several ways this can be applied. Consider the case where the loop gain constant, **K**, is changed. An example of this is if one were to inherit a PLL design done by someone else who has since left the company. Suppose in this case, the original design was done with a PLL with a charge pump gain of 1 mA, and now the PLL is being replaced with a newer one with a charge pump gain of 4 mA. Because the loop gain changes by a factor of 4, all the capacitors should be made four times larger and all the resistors should be made to one-fourth of their original value. Consider a second case where one designed a loop filter for a loop bandwidth of 10 kHz, but now wants to increase the loop bandwidth to 20 kHz. In this case, the loop bandwidth is changing by a factor of two, so the capacitors should be one-fourth of their original value, and the resistors should be twice their original value.

Scaling Rule of Thumb for the Loop Bandwidth

This rule deals with the case when the gain constant is changed, but the loop filter components are not changed and the impact on loop bandwidth is desired. Even though the loop filter is not optimized in this case, it still makes sense to understand how it behaves. An example of this could be where the loop filter is designed for a particular VCO gain, but the actual VCO has a gain that varies considerably.

To derive this rule, recall that the loop bandwidth is the frequency for which the magnitude of the open loop transfer function is unity. It is easier to see this relationship if the open loop transfer function is expressed in the following form.

$$G(s) = \frac{K}{s} \cdot \frac{1 + s \cdot T2}{s \cdot A0 \cdot (1 + s \cdot T1) \cdot (1 + s \cdot T3) \cdot (1 + s \cdot T4)} \quad (9.12)$$

Although it is a coarse approximation to neglect all the poles and zero, one can derive the fundamental elegant result by doing so.

$$Fc \propto \sqrt{K} \quad (9.13)$$

This rule assumes that the loop filter is not changed. For example, suppose that it is known that a PLL has a loop bandwidth of 10 kHz when the VCO gain is 20 MHz/V. Suppose the VCO gain is actually 40 MHz/V. In this case, the loop bandwidth should be about 1.4 times larger. This rule of thumb is not exact and the loop filter may not be perfectly optimized, but it is useful in many situations.

Conclusion

This chapter has discussed the fundamental concept of PLL transfer functions. The reader should familiarize themselves with the notation in this chapter, since it will be used throughout the book. Although the PLL transfer functions are derived as phase transfer functions, the frequency transfer functions are identical.

Chapter 10   Reference Spurs and their Causes

Introduction

In PLL frequency synthesis, reference sidebands and spurious outputs are an issue in design. There are several types of these spurious outputs with many different causes. However, by far, the most common type of spur is the reference spur. These spurs appear at multiples of the comparison frequency.

This chapter investigates the causes and behaviors of these reference spurs. In general, spurs are caused by either leakage or mismatch of the charge pump. Depending on the cause of the reference spurs, the spurs may behave differently when the comparison frequency or loop filter is changed. This chapter will discuss how to determine which is the dominant cause for a given application. In order to discuss spur levels, the fundamental concept of spur gain will be introduced. A clear understanding of spur gain is the starting point to understanding how reference spurs will vary from one filter to another. After this concept is developed, leakage and mismatch dominated spurs will be discussed, and then these results will be combined.

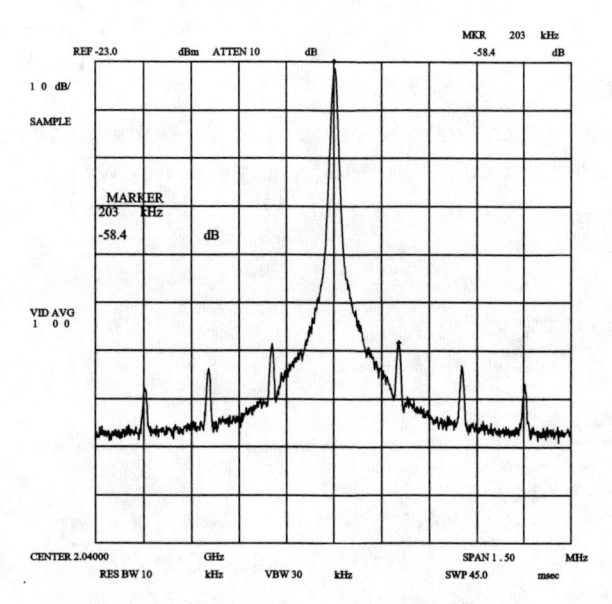

Figure 10.1 *Typical Reference Spur Plot*

The Definition of Spur Gain

Conceptually, if a given current noise of a fixed frequency is injected into the loop filter from the charge pump, then the power of the frequency noise that this induces at the VCO would be a start to defining the spur gain. An additional factor of *1/s* is included in the transfer function to simplify the arithmetic later. Note that since this is a frequency change, it is necessary to multiply the transfer function by a factor of *s* to convert from phase to frequency. This factor of *1/s* is left in, because it turns out that it is reintroduced because of other factors. Furthermore it makes the concept of spur gain a dimensionless quantity. Now since the power of the reference spur is sought, it is necessary to square this gain, and it is finally expressed in decibels for convenience.

$$\text{Spur Gain}(Fspur) = 20 \bullet \log |CL(s)|_{s=j \bullet Fspur \bullet 2 \bullet \pi} \tag{10.1}$$

So spur gain is the closed loop transfer function evaluated at the spur offset frequency of interest, **Fspur**. In most cases, **Fspur** will be assumed to be the comparison frequency, **Fcomp**, but it could also be other frequencies, such as multiples of the comparison frequency, or fractions of the comparison frequency (in the case of a fractional N PLL). In cases where the spur frequency of interest is outside the loop bandwidth, the spur gain can be approximated using the open loop transfer function instead of the closed loop transfer function. This greatly simplifies some of the mathematical analysis done later on.

The levels of reference spurs are directly related to the spur gain. In other words, if the spur gain decreases 1 dB, one would expect the spur at that frequency as well to decrease by 1 dB. The derivation of this is given in the Appendix and the approximations used hold very well provided that the spur level that is predicted is –10 dBc or less. Aside from spur gain, there are other factors that contribute to spur levels, depending on whether the spurs are leakage dominated or mismatch dominated.

Leakage Dominated Spurs

At lower comparison frequencies, leakage effects are the dominant cause of reference spurs. When the PLL is in the locked condition, the charge pump will generate short alternating pulses of current with long periods in between in which the charge pump is tri-stated.

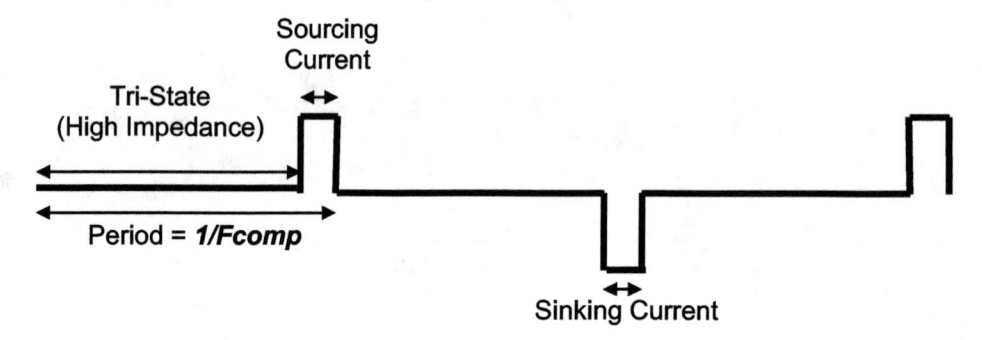

Figure 10.2 *Output of the Charge Pump When the PLL is in the Locked Condition*

When the charge pump is in the tri-state state, it is ideally high impedance. There will be some parasitic leakage through the charge pump, VCO, and loop filter capacitors. Of these leakage sources, the charge pump tends to be the dominant one. This causes FM modulation on the VCO tuning line, which in turn results in spurs. This is described in greater detail in the appendix. To predict the reference spur levels based on leakage, use the following general rule:

$$\text{LeakageSpur} = \text{BaseLeakageSpur} + 20 \bullet \log\left(\frac{\text{Leakage}}{K\phi}\right) + \text{SpurGain} \tag{10.2}$$

The leakage due to the PLL charge pump is temperature dependent and is often given guaranteed ratings as well as typical ratings and graphs in performance. The leakage of the charge pump increases with temperature, so spurs caused by leakage of the charge pump tend to increase when the PLL is heated. Various leakage currents were induced at various comparison frequencies, and the results were measured on the bench. The loop filter was not changed during any of these measurements. These results imply the fundamental constant for leakage-dominated spurs:

$$BaseLeakageSpur = 16.0 \ dBc \qquad (10.3)$$

Note that this constant is universal and not part specific and should apply to any integer PLL. It can also not be stressed enough that it is impossible to directly measure the *BaseLeakageSpur* – this number is extrapolated from other numbers.

I_{leak} (nA)	$20 \cdot Log(I_{leak}/K\phi)$ (dB)	$Fcomp$ (kHz)	Filter	Spur Levels (dBc)			Spur Gain (dB)			Implied BaseLeakage Spur (dBc)		
				1st	2nd	3rd	1st	2nd	3rd	1st	2nd	3rd
200	-86.0	50	A	-28.3	-40.5	-47.3	41.7	29.7	22.7	16.0	15.8	16.0
100	-92.0	50	A	-33.8	-45.7	-52.7	41.7	29.7	22.7	16.5	16.6	16.6
100	-80.0	100	B	-24.3	-40.5	-51.5	38.8	21.9	11.6	16.9	17.6	16.9
100	-80.0	200	B	-43.5	-61.5	-72.0	21.9	4.2	-6.3	14.6	14.3	14.3
500	-46.0	400	C	-32.7	X	X	-2.4	X	X	15.7	X	X
200	-54.0	400	C	-40.5	X	X	-2.4	X	X	15.9	X	X
Average Base Leakage spur										15.9	16.1	16.0
Filter	$K\phi$ (mA)	$Kvco$ (MHz/V)	C1 (nF)	C2 (nF)	C3 (pF)	R2 (KΩ)	R3 (KΩ)	Output Frequency (MHz)				
A	4.0	17	5.6	33	0	4.7	0	900				
B	1.0	43	0.47	3.3	90	12	39	1960				
C	0.1	48	1	4.7	0	18	0	870				

Figure 10.3 *Spur Level vs. Leakage Currents and Comparison Frequency*

Note that the *BaseLeakageSpur* index applies to the primary reference spurs as well as higher harmonics of this spur. Appendix B shows a theoretical calculation that derives this same result to textbook accuracy.

Pulse Related Spurs

In classical PLL literature, it is customary to model the reference spurs based entirely on leakage currents. For older PLLs, where the leakage currents were in the µA range, this made reasonable estimates for reference spurs and their behavior. However, modern PLLs typically have leakage currents of 1 nA or less, and therefore other factors tend to dominate the spurs, except at low comparison frequencies.

Recall that the charge pump comes on for very short periods of time and then is off during most of the time. It is the length of time that these short charge pump corrections are made that determines the pulse related spur. In other words, if leakage is not the dominant factor, then it is this time that the charge pump is on that determines the spur levels. There are several factors that influence this correction pulse width which include dead-zone elimination circuitry, charge pump mismatches, and unequal transistor turn on times.

The dead zone elimination circuitry forces the charge pump to turn on in order to keep the phase detector out of the dead zone. It is this period that the charge pump is on that is the root cause of reference spurs when charge pump leakage is not a factor. Note that even though leakage is not the cause of pulse related spurs, it can have a small influence on this pulse width.

Mismatch and unequal turn on times of the charge pump transistors also have a large impact on this minimum turn on time for the charge pump. When the charge pump source and sink currents are not equal, they are said to have mismatch. For instance, if the source current were 10% higher than the sink current, then a rough rule of thumb would be that the charge pump would have to come on 10% longer than its minimum on time when sinking current, producing an overall increase in spur levels. The unequal turn on times of the sink and source transistors also can increase this charge pump on time. In general, the source transistor is a PMOS device, which has twice the turn on time as the sink transistor, which is an NMOS device. The net effect of this is that the effective source current is reduced, and this has a similar effect as having negative mismatch. To illustrate this issue, consider the National Semiconductor LMX2315 PLL for which the optimal spur levels occur around +4% mismatch instead of 0% mismatch due to these unequal transistor turn on times.

For pulse related spur issues, it is important to be aware of the mismatch properties and to base the design around several different parts to get an idea of the full variations. Mismatch properties of parts can vary from date code to date code, so it is important to consider that in the design process. Also, in designs where an op-amp is used in the loop filter, it is best to center the op-amp around half of the charge pump supply voltage or slightly higher. Due to this variation of spur level over tuning voltage to the VCO, the way that spurs are characterized in this chapter are by the worst-case spur when the VCO tuning voltage is varied from 0.5 volts to 0.5 volts below the charge pump supply. The variation can also be mentioned, since this shows how much the spur varies, but ultimately, the worst-case spur should be the figure of merit. To predict reference spurs caused by the pulsing action of the charge pump, the following rule applies.

$$Pulse\ Spur = BasePulseSpur + Spur\ Gain + 40 \bullet \log\left(\frac{Fspur}{1\,Hz}\right) \qquad (10.4)$$

The reader may be surprised to see that the above formula has the additional *Fspur* term added. This was first discovered by making observations with a modulation domain analyzer, which displays frequency versus time. In the case of the leakage-dominated spur, the VCO frequency was assumed to be modulated in a sinusoidal manner, which was confirmed with observations on the bench. However, this was not the case for the pulse-dominated spur. For these, frequency spikes occur at regular intervals of time corresponding to when the charge pump turns on. The pulse-dominated spurs were measured and their magnitude could be directly correlated to the magnitude of these frequency spikes. This correlation was independent of the comparison frequency. Therefore, using the modulation index concept does not work for pulse dominated spurs and introduces an error equal to *20•log(Fspur)*. The pulse spur differs from the leakage spur not by this factor but by *40•log(Fspur)*. The additional factor of *20•log(Fspur)* comes because it is more proper to model the charge pump noise as a train of pulse functions, not a sinusoidal function. Recall to recover the time domain response of a pulse function applied to a system, this is simply the inverse Laplace transform. In a similar way that the inverse Laplace transform of *1/s* is just *1*, and not involving any factors of *1/ω*, likewise in this situation, a factor of *1/ω* is lost for this reason, thus accounting for the additional factor of *40•log(Fspur)*.

Fout MHz	N	Fspur kHz	Kϕ mA	Kvco MHz/V	C1 nF	C2 nF	C3 pF	R2 KΩ	R3 KΩ	Spur dBc	Spur Gain dB	BasePulse Spur dBc
This data was all taken from an LMX2330 PLL. The VCO was near the high end of the rail.												
1895	18950	100	4	43.2	2.2	10	0	6.8	0	-51.7	46	-297.7
1895	18950	100	4	43.2	13.9	66	0	2.7	0	-69.7	30	-299.7
1895	18950	100	4	43.2	0.56	2.7	0	15	0	-41.0	58	-299.0
1895	18950	100	4	43.2	1.5	6.8	0	5.6	0	-50.0	49.2	-299.2
1895	18950	100	4	43.2	1.5	6.8	100	5.6	39	-59.8	40.5	-300.3
1895	6064	312.5	4	43.2	4.7	20	0	1.8	0	-60.2	19.6	-299.6
1895	6064	3125.	4	43.2	1.8	5.6	0	1.5	0	-51.1	27.7	-298.6
This data was taken from an LMX2326 PLL with Vtune = 0.29 V and Vcc = 3 V												
231	1155	200	1	12	0.47	3.3	0	12	0	-74.1	23.0	-309.1
881.6	4408	200	1	18	0.47	3.3	0	12	0	-70.1	27.6	-309.7
881.6	1146	770	1	18	0.47	3.3	0	12	0	-70.1	4.9	-308.8
1885	9425	200	1	50	0.47	3.3	0	12	0	-59.7	35.6	-308.6
1885	4343	434	1	12	0.47	3.3	0	12	0	-58.7	22.2	-307.7

Table 10.1 *Demonstration of the Consistency of the BasePulseSpur*

The first several rows in Table 10.1 demonstrate many different filters at the same output frequency. The last several rows use the same filter, but emphasize the difference in changing the *N* value and comparison frequency. For the last several rows, the charge pump voltage was kept at 0.29 volts to maintain consistent mismatch properties of the charge pump and to also make spurs that were easy to measure. For this reason, this table is a valuable tool to show how spur levels vary. However, it is not a good source of information for worst-case **BasePulseSpur**, since the tuning voltage was within 0.5 V of the supply rail and therefore out of specification.

PLL	Variation (dBc)	BasePulseSpur (dBc)
LMX2301/05, LMX2315/20/25	11	-299
LMX2330/31/32/35/36/37	23	-311
LMX2306/16/26	7	-309
LMX1600/01/02	5	-292
LMX2430/33/34, LMX2470 IF	5	-333

Table 10.2 *BasePulseSpur for Various National Semiconductor PLLs*

Despite the tables and measurements given above, the avid reader is sure to try to relate the pulse related spur to the mismatch of the charge pump. To do this, the LMX2315 PLL was used, and the spur level was measured along with the charge pump mismatch. The spur gain of this system was 19.6 dB, and in this system the comparison frequency was 200 kHz, so the spurs are clearly pulse-dominated. Note that the turn-on time of the charge pump transistors also comes into play, so this result is specific to the LMX2315 family of PLLs.

Vtune (Volts)	1	1.5	2.2	3	4	4.5
Source (mA)	5.099	5.169	5.241	5.308	5.397	5.455
Sink (mA)	5.308	5.253	5.166	5.047	4.828	4.517
mismatch (%)	- 4.0	- 1.6	1.4	5.0	11.1	18.8
200 kHz Spur (dBc)	- 73.1	- 76.6	- 83.3	- 83.2	- 72.8	- 65.7

Table 10.3 *Sample Variation of Spur Levels and Mismatch with Do voltage*

Using statistical models, this suggests that the best spur performance is actually when the charge pump is 3.2 % mismatched and also gives the relationship:

$$BasePulseSpur = -315.6 + 1.28 \bullet | \%mismatch - 3.2\% | \qquad (10.5)$$

Combining the Concepts of Leakage Related Spurs and Pulse Related Spurs

Critical Values for Comparison Frequency

In most cases, it makes sense to model the spurs as pulse related spurs, but this may not work for low comparison frequencies. One way to determine if a spur is leakage or pulse related is to calculate spurs based on both methods, and use whichever method yields the largest spur levels. In most cases, the pulse related spur will dominate. If the leakage is known, and the *BasePulseSpur* is known, it is possible to predict the comparison frequency for which the spur is equally pulse and leakage dominated. If the comparison frequency is higher than this, then the spur becomes more pulse dominated. Note that this calculation is independent of the spur gain and is found by setting the leakage spur equal to the pulse spur and solving for the comparison frequency. Equation (10.6) and Table 10.4 show how to calculate the critical frequencies and give some values for reference.

$$40 \cdot \log\left(\frac{Fcomp}{1Hz}\right) = (BaseLeakageSpur - BasePulseSpur) + 20 \cdot \log\left(\frac{leakage}{K\phi}\right) \quad (10.6)$$

Comparison frequencies that satisfy this equation will be called critical frequencies. At the critical frequency, the reference spur is equally dominated by leakage and pulse effects. Above the critical frequency, the spur becomes more pulse dominated, below the critical frequency, the spur becomes more leakage dominated. This table was generated assuming the a charge pump gain of 1 mA and a **BaseLeakageSpur** of 16.0 dBc. Note that the critical frequency is proportional to the square root of the leakage current, and inversely proportional to the square root of the charge pump gain.

		\multicolumn{7}{c}{*BasePulseSpur (dBc)*}
		-290	-300	-310	-320	-330	-340	-350
Leakage (nA)	0.1	14.1	25.1	44.7	79.4	141.3	251.2	446.7
	1.0	44.7	79.4	141.3	251.2	446.7	794.3	1412.5
	10.0	141.3	251.2	446.7	794.3	1412.5	2511.9	4466.8
	100.0	446.7	794.3	1412.5	2511.9	4466.8	7943.3	14125.4
	1000.0	1412.5	2511.9	4466.8	7943.3	14125.4	25118.9	44668.4

Table 10.4 *Critical Values for Comparison Frequency in Kilohertz*

Composite Spur Calculation

This chapter has independently derived the spur levels based on leakage and pulse effects. Regardless of the dominant cause, the spur level is given by:

$$Spur = 10 \cdot \log\left(10^{LeakageSpur/10} + 10^{PulseSpur/10}\right) \quad (10.7)$$

Spur Levels vs. Unoptimized Loop Filter Parameters

Using the expression for spur gain, the way that spur levels vary vs. various parameters can easily be calculated and is shown below:

Parameter	Description	Leakage Dominated Spurs	Pulse Dominated Spurs
i_{leak}	Charge Pump Leakage,	$20 \cdot \log(i_{leak})$	N/A
M	Charge Pump	N/A	Correlated to $\lvert M - \text{Constant} \rvert$
N	N Counter Value	independent	independent
Kvco	VCO Gain	$20 \cdot \log(Kvco)$	$20 \cdot \log(Kvco)$
Fcomp	Comparison Frequency	$-40 \cdot \log(Fcomp)$	$-20 \cdot \log(Fcomp)$
r	= Fcomp/Fc	$-40 \cdot \log(r)$	$-40 \cdot \log(r) + 20 \cdot \log(Fcomp)$
$K\phi$	Charge Pump Gain,	independent	$20 \cdot \log(K\phi)$
SG	Spur Gain	SG	SG

Table 10.5 *Spur Levels vs. Parameters if Loop Filter is NOT Redesigned*

Harmonics of Pulse Dominated Reference Spurs

In the case of a leakage-dominated spur, ***BaseLeakageSpur*** also applies to the spur harmonics, so this topic has already been covered. The case of pulse dominated spurs still needs to be discussed. In order to address this issue, a LMX2326 PLL was tuned in 1 MHz increments from 1900 MHz to 1994 MHz using an automated test program. For these tests, $K\phi$ = 1 mA, ***Fcomp*** = 200 kHz, and ***Kvco*** = 45 MHz/V. Filter A had components of ***C1*** = 145 pF, ***C2*** = 680 pF, ***R2*** = 33 KΩ, while Filter B had components of ***C1*** = 315 pF, ***C2*** = 1.8 nF, and ***R2*** = 18 KΩ.

	Fundamental (200 kHz)	2nd Harmonic (400 kHz)	3rd Harmonic (600 kHz)
Minimum (dBc)	-56.2	-65.1	-64.5
Average (dBc)	-52.8	-58.5	-61.9
Maximum (dBc)	-49.3	-54.4	-59.0
Spur Gain for Spur (dB)	45.7	33.8	26.8
BasePulseSpur (dBc)	-307.0	-312.4	-316.9

Table 10.6 *Reference Spurs and their Harmonics for Filter A*

	Fundamental (200 kHz)	2nd Harmonic (400 kHz)	3rd Harmonic (600 kHz)
Minimum (dBc)	-64.8	-70.4	-69.1
Average (dBc)	-60.8	-65.1	-66.8
Maximum (dBc)	-56.2	-61.1	-64.7
Spur Gain for Spur (dB)	39.0	27.1	20.0
BasePulseSpur (dBc)	-307.2	-312.2	-315.8

Table 10.7 *Reference Spurs and their Harmonics for Filter B*

Table 10.6 and Table 10.7 show that the pulse spur is relatively consistent for different filters, however the second harmonic has a different BasePulseSpur than the first. These empirical measurements would suggest to expect that the BasePulseSpur for the second harmonic to be about 5 dB better than the BasePulseSpur for the first harmonic, and for the BasePulseSpur of the third harmonic to be about 4 dB better than the BasePulseSpur for the second harmonic.

Table 10.6 and Table 10.7 show harmonics of pulse dominated reference spurs. Similar measurements can also be made for harmonics of leakage-dominated spurs. Theoretically, one would expect that the higher harmonics to behave differently than the fundamental leakage dominated spur, since they are based on the higher powers of the modulation index (See Appendix A), however measured results show that they can be treated just as the fundamental leakage spur, except for the value of ***BaseLeakageSpur*** for them is a little different.

	Fundamental (200 kHz)	2nd Harmonic (400 kHz)	3rd Harmonic (600 kHz)
Minimum (dBc)	-56.2	-65.1	-64.5
Average (dBc)	-52.8	-58.5	-61.9
Maximum (dBc)	-49.3	-54.4	-59.0
Spur Gain for Spur (dB)	45.7	33.8	26.8
BasePulseSpur (dBc)	-307.0	-312.4	-316.9

Table 10.8 *Reference Spurs and their Harmonics for Filter A*

Conclusion

This chapter has discussed the causes of reference spurs and given some techniques to simulate their general behavior. The concept of spur gain applies to reference spurs and gives a relative indication of how they vary from one loop filter to another when the other parameters, such as comparison frequency are held constant. Reference spurs can be caused by leakage or pulse effects. Pulse effects is a generic term to refer to inconsistencies in the pulse width of the charge pump caused by mismatch, or unequal transistor turn on times. Although reference spurs are intended to refer to spurs that appear at a spacing equal to the comparison frequency from the carrier, the models in this chapter are also useful in predicting harmonics of reference spurs and fractional spurs. One caution dealing with fractional spurs is they may be sensitive to voltage and prescaler. They also often have a dependence on the output frequency as well. In general, the spur that is closest to the carrier is the most troublesome, since it is most difficult to filter.

As for the accuracy of the formulas presented in this chapter, there will always be some variation between the actual measured result and the theoretical results. Relative comparisons using spur gain tend to be the most accurate. It is recommended to use the empirical value, but to accept that there could be several dB variation between the predicted and measured results. In the case of pulse-dominated spurs, the value for *BasePulseSpur* is purely empirical and is based solely on measured data. These spurs can also change a good 15 dB as the VCO is tuned across its tuning range. It is the worst-case spur is the one that is being modeled.

Appendix A: Spectra of Spurious Signals

Introduction

This section investigates the causes of spurs and their spectral density for an arbitrary time-varying signal that is fed to a VCO. It assumes a sinusoidal signal and is therefore meaningful in analyzing leakage-dominated spurs.

Derivation of Spurious Spectrum

Spurs are caused by the PLL when a signal with an AC component is presented to the tuning line of the VCO. Assume that the tuning voltage to the VCO has the form:

$$V_{tune} = V_{DC} + V_{AC}(t) \tag{10.8}$$

Where
- V_{tune} = Tuning voltage to the VCO
- V_{DC} = DC component of tuning voltage to the VCO
- V_{AC} = AC component of tuning voltage to the VCO
- $\quad\quad = V_m \cdot sin(\omega_m \cdot t)$
- ω_m = Modulating Frequency = **Fcomp**

The VCO has an output voltage of the form (Tranter 1985):

$$V(t) = A \cdot cos[\omega_0 \cdot t + \beta \cdot sin(\omega_m \cdot t)] \tag{10.9}$$

Where
- ω_0 = Carrier Frequency
- β = Modulation Index

Since $\beta \cdot sin(\omega_n \cdot t)$, represents the phase deviation of the signal, this expression can be differentiated to determine the frequency deviation, ΔF, and the following identity can be derived (Tranter 1985):

$$\beta = \frac{\Delta F}{\omega_n} \tag{10.10}$$

Now $e^{j \cdot \beta \cdot sin(\omega_n \cdot t)}$ can be expanded in a complex exponential Fourier series as follows (Tranter 1985):

$$e^{j \cdot \beta \cdot sin(\omega_n \cdot t)} = \sum_{n=-\infty}^{\infty} J_n(\beta) \cdot e^{j \cdot n \cdot \omega_m \cdot t} \tag{10.11}$$

In the above expression, $J_n(\beta)$ is the Bessel function of the first kind of order n.

Applying the identity allows the power spectral density to be simplified as follows (Tranter 1985):

$$Vout(t) = A \cdot \cos[\omega_0 \cdot t + \beta \cdot \sin(\omega_m \cdot t)] \quad (10.12)$$

$$= A \cdot Real\left\{ e^{j \cdot \omega_0 \cdot t} \sum_{n=-\infty}^{\infty} J_n(\beta) \cdot e^{j \cdot n \cdot \omega_m \cdot t} \right\}$$

$$= A \cdot \sum_{n=-\infty}^{\infty} J_n(\beta) \cdot \cos(\omega_0 \cdot t + n \cdot \omega_m \cdot t)$$

From this expression, the sideband levels can be found by visual inspection.

$$\begin{aligned} Carrier &: J_0(\beta) \approx 1 \\ First &: J_1(\beta) \approx \frac{\beta}{2} \\ Second &: J_2(\beta) \approx \frac{\beta^2}{8} \\ n^{th} &: J_n(\beta) \end{aligned} \quad (10.13)$$

Below is a table of first sideband level versus frequency deviation from zero for various comparison frequencies:

Spur Level (dBm)	Modulation Index (β)	Frequency Deviation for Various Comparison Frequencies (Hz)					
		Fcomp 10 kHz	Fcomp 30 kHz	Fcomp 50 kHz	Fcomp 100 kHz	Fcomp 200 kHz	Fcomp 1000 kHz
-30	6.32e-2	632	1900	3160	6320	12600	63200
-40	2.00e-2	200	600	1000	2000	4000	20000
-50	6.32e-3	63	190	316	632	1260	6320
-55	3.56e-3	36	107	178	356	712	3560
-60	2.00e-3	20	60	100	200	400	2000
-65	1.12e-3	11	34	56	112	224	1120
-70	6.32e-4	6	19	32	63	126	632
-75	3.56e-4	4	11	18	36	71	356
-80	2.00e-4	2	6	10	20	40	200
-85	1.12e-4	1	3	6	11	22	112
-90	6.32e-5	0.6	2	3	6	13	63

Table 10.9 *Spur Level, Modulation Index, and Frequency Variation*

The spur levels relate the modulation index by:

$$\text{Spur Level} = 20 \cdot \log(\beta/2) \quad\quad (10.14)$$

Bessel Correction for Spur Gain

By using the approximation to the first order Bessel Function shown in equation (10.13), one can derive equation (10.14), which implies that if the spur gain changes by 1 dB, then so do the spurs. In actuality, this disregards the higher order terms in the Taylor Series for the Bessel function, which can be relevant in some cases. The error in this assumption will be called the Bessel Corection and is shown in Figure 10.4 . For almost all situations, neglecting the Bessel Correction makes an excellent prediction and the calculation error is far less than any measurement error. If the spur level is –10 dBc or lower, the error is about –0.15 db, and the magnitude of this error gets much smaller at a rate of 10 dB/decade as the spur level gets lower. So provided that the spur level that is theoretically calculated is less than –10 dBc, there is no reason to worry about this correction. In the case of fractional PLLs, this factor is disregarded, but there may be some obscure cases where it may become relevant. In order to simplify calculations, this correction factor will be assumed to be negligible and will be disregarded in calculations for the rest of this book. Figure 10.4 shows the impact of correcting the spur level with the Bessel Correction.

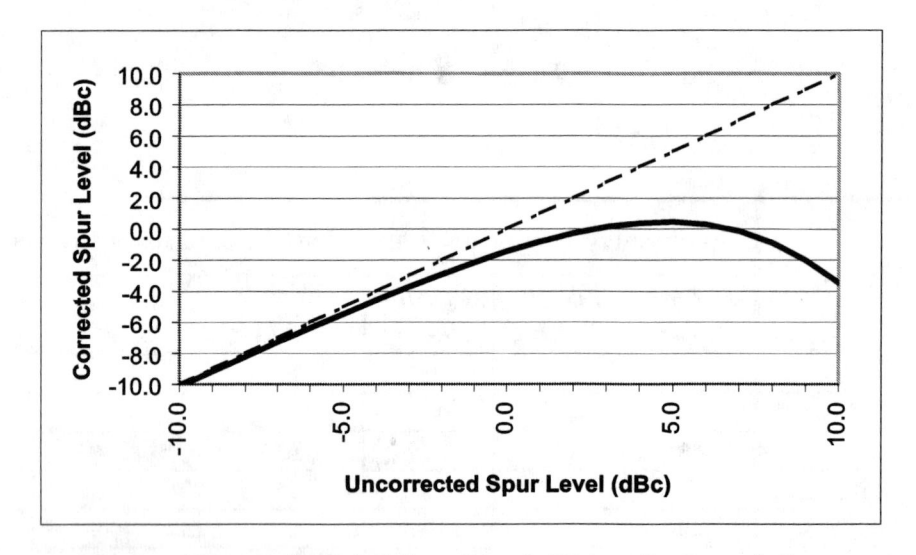

Figure 10.4 *Calculated Spur Levels Using the Bessel Correction*

References

Tranter, W.H. and R.E. Ziemer *Principles of Communications Systems, Modulation, and Noise*, 2nd ed, Houghton Mifflin Company, 1985

Appendix B: Theoretical Calculation of Leakage Based Spurs

Since the ***BaseLeakageSpur*** is theoretically independent of PLL and loop filter, it makes sense to choose the loop filter that is the most basic. A simple capacitor is the most basic loop filter. Although this filter topology is not a stable one, it us sufficient for the purposes of calculations. Using this simplified loop filter, the voltage deviation to the VCO can easily be calculated.

$$\frac{\Delta V}{\Delta t} = \frac{i}{C1} \quad (10.15)$$

Substituting in known values gives the voltage deviation.

$$\Delta V = \int_0^{1/Fcomp} \frac{i}{C1} \cdot dt = \frac{i}{C1 \cdot Fcomp} \quad (10.16)$$

Now recall that this is the amount the voltage changes during one charge pump cycle. So to get the modulation index, it is necessary to divide by two. Therefore, the modulation index is:

$$\beta = \frac{Kvco \cdot \Delta V}{2 \cdot Fcomp} \quad (10.17)$$

$$Leakage\ Spur = 20 \cdot \log\left(\frac{\beta}{2}\right) \quad (10.18)$$

Table 10.10 shows the fundamental result and how it can be derived. This number is within a few dB of what has been measured in practice.

Specified Quantities		Derived Quantities	
C1	10 nF	**Spur Gain**	8.073 dB
C2	0 nF	ΔV	1.000 µV
R2	0 KΩ	β = modulation index	0.00005
Kφ	1 mA	**Leakage Spur**	= 20•log(β/2) = -92.041 dBc
Kvco	10 MHz/V	20•(Leakage/Kφ)	-120.000
Leakage	1 nA	**BaseLeakageSpur**	= -92.0 dBc – (-120 dB) – 8.073 dB = **19.886 dBc**
Fcomp	100 kHz		

Table 10.10 *Theoretical calculation for BaseLeakageSpur = 19.9 dBc/Hz*

Correction for Base Leakage Spur

Now the above calculations derive a number for *BaseLeakageSpur* of 19.9 dBc. However, it is more accurate to model the modulation waveform as a triangle wave instead of a sine wave. Doing this gives a result that agrees very closely with the measured results.

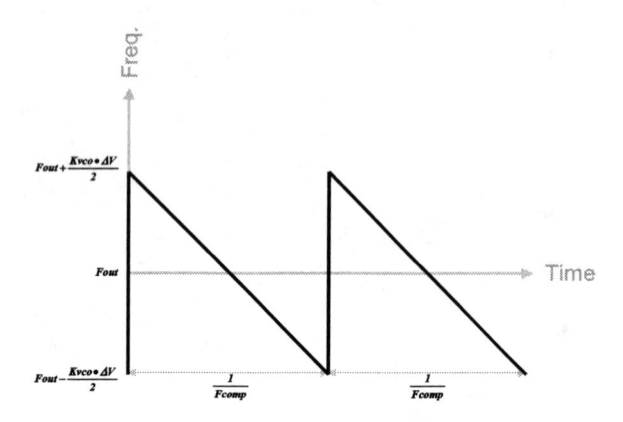

Figure 10.5 *VCO Frequency Output*

If one compares the first Fourier coefficient of the right triangle wave to that of a sine wave of the same amplitude, it is found that the magnitude of the right triangle is slightly less. The sine wave coefficient has a magnitude of $2/\pi$, while the sine wave has a coefficient of 1. The correction factor is therefore:

$$20 \bullet log\left(\frac{2}{\pi}\right) = -3.9223 \qquad (10.19)$$

Adding in this factor gives the fundamental result that:

$$BaseLeakageSpur = 15.96\,dBc \qquad (10.20)$$

Chapter 11 Fractional Spurs and their Causes

Introduction

In fractional PLLs, spurs appear at a spacing equal to the channel spacing, even though the comparison frequency is much higher. These spurs that appear at the channel spacing in fractional PLLs are traditional fractional spurs. There are different ways of compensating for these spurs, each way having its own rules and exceptions. The easiest spur to understand and predict is the uncompensated fractional spur. After this, the impact of compensation can be discussed with better understanding. Note that although higher order delta sigma PLLs do not have analog compensation, delta sigma modulation will be treated as a form of compensation, because it introduces many of the exceptions and special rules that analog compensation does. There are the concepts of in-band spurs and rolloff that are useful tools in understanding fractional spurs.

In-Band Spurs

One unique property of fractional PLLs over integer PLLs is that it is possible to have spurs within the loop bandwidth of the PLL and still have a stable system. Even though it is not feasible in many applications to have this, it still provides a useful understanding of fractional spurs and an understanding of what the fractional spurs will look like outside the loop bandwidth. On some PLLs, it is possible to disable the fractional compensation. The effects of these uncompensated spurs were studied at different VCO frequencies, comparison frequencies, charge pump gains, and VCO gains. It was found that the in-band spurs that were calculated were independent of all these factors. For uncompensated fractional spurs, the spurs may become higher than the carrier when brought inside the loop bandwidth. For this reason, these in-band uncompensated fractional spurs were not measured directly, but rather calculated by measuring fractional spurs outside the loop bandwidth and then accounting for how much the loop filter rolled off the spurs.

Rolloff

The amount of fractional spur supression that a loop filter gives is defined as rolloff and is calculated as shown below:

$$rolloff = Spur\ Gain\ -\ 20\bullet log(N) \qquad (11.1)$$

In addition to being easy to calculate directly, rolloff can be approximated using a spectrum analyzer, provided that the VCO noise does not impact the measurement. In order to measure it, one measures the phase noise at the frequency of interest and then subtracts away the in-band phase noise. Although it can be approximated with a measurement, it is always more accurate to explicitly calculate it.

Mathematical Calculation of Uncompensated Fractional Spurs

Underlying Theory

The concept of modeling fractional spurs is to assume that the loop bandwidth is infinite and consider the resulting spectrum at the VCO output. Now the VCO output will be toggling between two frequency values. The output of the VCO can be expressed as follows:

$$f(t) = m(t) \bullet \cos(2\pi \bullet f1 \bullet t) + (1 - m(t)) \bullet \cos(2\pi \bullet f2 \bullet t) \qquad (11.2)$$

In this case, *m(t)* is the modulating signal. This has a value of either zero or one and corresponds is the overflow output of the fractional accumulator. In order to find the spectrum, we take the Fourier transform of the above expression. Recall that the Fourier transform a sum is the sum of the Fourier transforms and that the Fourier transform of a product is the convolution of the Fourier transforms. Applying these identities yields the following relationship.

$$\begin{aligned}\Im\{f(t)\} &= M(s) \otimes \left(\frac{\delta(s - 2\pi \bullet f1) + \delta(s + 2\pi \bullet f1)}{2}\right) \\ &\quad + (\delta(s) - M(s)) \otimes \left(\frac{\delta(s - 2\pi \bullet f2) + \delta(s + 2\pi \bullet f2)}{2}\right) \\ &= \frac{M(s - 2\pi \bullet f1) + M(s + 2\pi \bullet f1)}{2} - \frac{M(s - 2\pi \bullet f2) + M(s + 2\pi \bullet f2)}{2}\end{aligned} \qquad (11.3)$$

What the above equation demonstrates is that the spectrum of PLL output will be the spectrum of the modulating signal shifted in frequency. Therefore, in order to predict the fractional spurs, all that is necessary is to expand the modulating signal, *m(t)*, in a Fourier series. Note that in general, the modulating signal is neither even nor odd, so both the cosine and sine terms are necessary. By taking 20•Log of the magnitude of these, the spur levels can be calculated.

Fractional Spur Symmetry

After deriving the above equations, one will quickly realize the symmetry property of fractional spurs. This means that if one chooses a fractional numerator of one or one less than the fractional denominator (**FDEN**), the spurs are the same. This basically has the impact of switching f1 and f2. So in general, the spur for a fractional numerator of **FNUM** and **FDEN** minus **FNUM** will be the same. For instance, a fraction of 1/16 and 15/16 yield the same fractional spur spectrum.

Fractional Spur Chart

Spur Order	1	2	3	4	5	6	7	8	9	10	11	12	13	14	15	16	17	18	19	20
Spur Power	0.0	-6.0	-9.5	-12.0	-14.0	-15.6	-16.9	-18.1	-19.1	-20.0	-20.8	-21.6	-22.3	-22.9	-23.5	-24.1	-24.6	-25.1	-25.6	-26.0

Table 11.1 *Fractional Spur Chart*

The result of performing the fractional spur simulations is the above chart. The top row of the chart is the spur order. By this, it is meant that 1 is the worst-case spur, 2 is the second worse case spur, and so on. In this case, it is understood that fractional spurs come in pairs at *FNUM* and *FDEN-FNUM*, and it is these pairs that are being counted in this table. On the bottom row is the spur level in dBc. So the worst-case in-band spur is always 0 dBc. Note that this applies to any fractional spur be it the first, second, or whatever. The next question becomes how one determines what fractional numerator produces the worst-case spur. This is a function of the spur order. This is best illustrated by an example. Consider a fractional denominator of 100.

First Fractional Spur

The first thing to do is generate the sequence:

$$1, \left\lfloor \frac{FDEN}{2} \right\rfloor, \left\lfloor \frac{FDEN}{3} \right\rfloor, \left\lfloor \frac{FDEN}{4} \right\rfloor, \left\lfloor \frac{FDEN}{5} \right\rfloor \qquad (11.4)$$

In this case, this turns out to be 1, 50, 33, 25, 20. Now the first element and its complement are the worst-case numerators. So 1 and 99 are the worst-case numerators. Then on the spur chart, we see the worst-case in-band spur is 0 dBc.

Now proceed to the next numerator. In this case, the number is 50. Now because this has a factor in common with the fractional denominator, there is no first fractional spur here and this one will be skipped. Had the fractional denominator been 101, then there would be a spur for this numerator of level –6 dBc. Even though there was no spur present for this numerator, it is still counted, and we progress down the chart for the next spur. The next number in the sequence is 33. This and the complementary frequency of 67 produce an in-band spur of –9.5 dBc. The next number is 25. Because this has a factor in common with 100, there is no spur here. The next number is 20, which also has a factor in common with 100, so there is no spur here either. Now the exception comes when the same number is repeated in the sequence. In this case, the spur power is correct, but the numerator is slightly shifted. Describing the generation of this sequence is tedious, but it is recommended for the reader to use the table for the first fractional spur as both a reference and for better understanding of how to generate the numbers for the in-band spur.

Second and Higher Order Fractional Spur

For the second fractional spur, the most important thing to remember that the worst-case spur occurs for a fractional numerator of *2* and *FDEN* – *2*. In most cases, the second worst-case spur is at numerator of *1* and *FDEN-1*, but there are exceptions. In this case, it is probably best to use the table. For the kth fractional spur, the worst-case spurs occur when the fractional numerator is *k* or *FDEN-k*.

Uncompensated In-Band Fractional Spur Tables

The next several pages show simulations for fractional spur levels and numerators. There are definite patterns in the numbers. The power levels from the spur chart keep on appearing and the fractional numerators appear in accordance to the rules discussed so far. Note for very small denominator values, the power levels get distorted a little bit. Even up to a fractional denominator of about 30, these power levels are slightly different from the power level chart.

Here is an example of how to use this table. Consider a comparison frequency of 13 MHz and a fractional denominator of 65. The first fractional spur will be at 200 kHz offset from the carrier and the worst-case will occur at numerators 1 and 64 with a level of 0 dBc. The second worst-case will be for numerators 32 and 33, the level will be –6.0 dBc. The third worst-case will be at levels –9.5 dBc for fractional numerators of 22 and 43.

If one was interested in the second fractional spur at 400 kHz offset, the worst-case would be 0 dBc for fractional numerators of 2 and 61. The second fractional spur will be at 400 kHz offset and will occur at numerators 2 and 63. The second worst-case would be for numerators of 1 and 64 with a level of –6.0 dBc. The third worst-case would be for numerators of 31 and 34 with levels of –9.5 dBc.

Fractional Denominator	In-Band Fractional Spur				Fractional Numerator			
	Worst-case	2nd Worst	3rd Worst	4th Worst	Worst	2nd Worst	3rd Worst	4th Worst
2	-3.9	x	x	x	1	x	x	x
3	-1.6	x	x	x	1	x	x	x
4	-0.9	x	x	x	1	x	x	2
5	-0.6	-4.8	x	x	1	2	x	x
6	-0.4	x	x	x	1	x	x	3
7	-0.3	-5.4	-7.3	x	1	3	2	x
8	-0.2	-7.9	x	x	1	3	x	x
9	-0.2	-5.7	-9.4	x	1	4	2	x
10	-0.1	-8.5	x	x	1	3	x	x
11	-0.1	-5.8	-8.7	-10.3	1	5	4	3
12	-0.1	-11.5	x	x	1	5	x	x
13	-0.1	-5.8	-8.9	-10.8	1	6	4	3
14	-0.1	-9.0	-12.2	x	1	5	3	x
15	-0.1	-5.9	-11.1	-13.7	1	7	4	2
16	-0.1	-9.1	-12.6	-14.1	1	5	3	7
17	0.0	-5.9	-9.2	-11.3	1	8	6	4
18	0.0	-12.9	-14.7	x	1	7	5	x
19	0.0	-5.9	-9.3	-11.5	1	9	6	5
20	0.0	-9.3	-15.1	-16.0	1	7	3	9
21	0.0	-6.0	-11.6	-13.2	1	10	5	4
22	0.0	-9.3	-13.3	-15.5	1	7	9	3
23	0.0	-6.0	-9.4	-11.7	1	11	8	6
24	0.0	-13.4	-15.7	-17.6	1	5	7	11
25	0.0	-6.0	-9.4	-11.7	1	12	8	6
26	0.0	-9.4	-13.5	-15.9	1	9	5	11
27	0.0	-6.0	-11.8	-13.5	1	13	7	11
28	0.0	-9.4	-13.6	-17.6	1	9	11	3
29	0.0	-6.0	-9.4	-11.8	1	14	10	7
30	0.0	-16.1	-18.8	-19.4	1	13	11	7
31	0.0	-6.0	-9.4	-11.8	1	15	10	8
32	0.0	-9.4	-13.7	-16.2	1	11	13	9
33	0.0	-6.0	-11.9	-13.7	1	16	8	13
34	0.0	-9.5	-13.7	-16.3	1	11	7	5
35	0.0	-6.0	-9.5	-11.9	1	17	12	9
36	0.0	-13.7	-16.4	-19.5	1	7	5	13
37	0.0	-6.0	-9.5	-11.9	1	18	12	9
38	0.0	-9.5	-13.8	-16.4	1	13	15	11
39	0.0	-6.0	-11.9	-13.8	1	19	10	8
40	0.0	-9.5	-16.5	-18.4	1	13	17	9
41	0.0	-6.0	-9.5	-11.9	1	20	14	10
42	0.0	-13.8	-19.8	-20.9	1	17	19	13
43	0.0	-6.0	-9.5	-11.9	1	21	14	11
44	0.0	-9.5	-13.8	-16.6	1	15	9	19
45	0.0	-6.0	-11.9	-16.6	1	22	11	13
46	0.0	-9.5	-13.8	-16.6	1	15	9	13
47	0.0	-6.0	-9.5	-12.0	1	23	16	12
48	0.0	-13.8	-16.6	-20.1	1	19	7	13
49	0.0	-6.0	-9.5	-12.0	1	24	16	12
50	0.0	-9.5	-16.6	-18.6	1	17	7	11
51	0.0	-6.0	-12.0	-13.9	1	25	13	10
52	0.0	-9.5	-13.9	-16.7	1	17	21	15
53	0.0	-6.0	-9.5	-12.0	1	26	18	13
54	0.0	-13.9	-16.7	-20.2	1	11	23	5
55	0.0	-6.0	-9.5	-12.0	1	27	18	14
56	0.0	-9.5	-13.9	-18.7	1	19	11	25
57	0.0	-6.0	-12.0	-13.9	1	28	14	23
58	0.0	-9.5	-13.9	-16.7	1	19	23	25
59	0.0	-6.0	-9.5	-12.0	1	29	20	15
60	0.0	-16.7	-20.4	-21.6	1	17	11	23
61	0.0	-6.0	-9.5	-12.0	1	30	20	15
62	0.0	-9.5	-13.9	-16.7	1	21	25	9
63	0.0	-6.0	-12.0	-13.9	1	31	16	25
64	0.0	-9.5	-13.9	-16.7	1	21	13	9
65	0.0	-6.0	-9.5	-12.0	1	32	22	16
66	0.0	-13.9	-16.7	-21.7	1	13	19	5
67	0.0	-6.0	-9.5	-12.0	1	33	22	17
68	0.0	-9.5	-13.9	-16.8	1	23	27	29
69	0.0	-6.0	-12.0	-13.9	1	34	17	14
70	0.0	-9.5	-18.9	-20.5	1	23	31	19
71	0.0	-6.0	-9.5	-12.0	1	35	24	18
72	0.0	-13.9	-16.8	-20.5	1	29	31	13
73	0.0	-6.0	-9.5	-12.0	1	36	24	18
74	0.0	-9.5	-13.9	-16.8	1	25	15	21
75	0.0	-6.0	-12.0	-16.8	1	37	19	32
76	0.0	-9.5	-13.9	-16.8	1	25	15	11
77	0.0	-6.0	-9.5	-12.0	1	38	26	19
78	0.0	-13.9	-16.8	-20.5	1	31	11	7
79	0.0	-6.0	-9.5	-12.0	1	39	26	20
80	0.0	-9.5	-16.8	-18.9	1	27	23	9
81	0.0	-6.0	-12.0	-13.9	1	40	20	16
82	0.0	-9.5	-13.9	-16.8	1	27	33	35
83	0.0	-6.0	-9.5	-12.0	1	41	28	21
84	0.0	-13.9	-20.6	-21.9	1	17	23	13
85	0.0	-6.0	-9.5	-12.0	1	42	28	21
86	0.0	-9.5	-13.9	-16.8	1	29	17	37
87	0.0	-6.0	-12.0	-13.9	1	43	22	35
88	0.0	-9.5	-13.9	-16.8	1	29	35	25
89	0.0	-6.0	-9.5	-12.0	1	44	30	22
90	0.0	-16.8	-20.6	-22.0	1	13	41	7
91	0.0	-6.0	-9.5	-12.0	1	45	30	23
92	0.0	-9.5	-13.9	-16.8	1	31	37	13
93	0.0	-6.0	-12.0	-13.9	1	46	23	37
94	0.0	-9.5	-13.9	-16.8	1	31	19	27
95	0.0	-6.0	-9.5	-12.0	1	47	32	24
96	0.0	-13.9	-16.8	-20.6	1	19	41	35
97	0.0	-6.0	-9.5	-12.0	1	48	32	24
98	0.0	-9.5	-13.9	-19.0	1	33	39	11
99	0.0	-6.0	-12.0	-13.9	1	49	25	20
100	0.0	-9.5	-16.8	-19.0	1	33	43	11
128	0.0	-9.5	-14.0	-16.9	1	43	51	55
1920	0.0	-16.9	-20.8	-22.3	1	823	349	443
1968	0.0	-14.0	-16.9	-20.8	1	787	281	179

Table 11.2 *Calculated First Fractional Spur*

Fractional Denominator	In-Band Fractional Spur				Fractional Numerator			
	Worst-case	2nd Worst	3rd Worst	4th Worst	Worst	2nd Worst	3rd Worst	4th Worst
2	x	x	x	x	x	x	1	x
3	-13.7	x	x	x	1	x	x	x
4	-3.9	-9.9	x	x	2	1	x	x
5	-4.3	-8.4	x	x	2	1	x	x
6	-1.6	-7.7	x	x	2	1	x	x
7	-2.1	-7.2	-9.1	x	2	1	3	x
8	-0.9	-6.9	-6.9	x	2	1	3	x
9	-1.3	-6.7	-10.4	x	2	1	4	x
10	-0.6	-4.8	-6.6	-10.8	2	4	1	3
11	-0.8	-6.5	-9.4	-11.0	2	1	3	5
12	-0.4	-6.4	-6.4	x	2	1	5	x
13	-0.6	-6.4	-9.4	-11.3	2	1	5	6
14	-0.3	-5.4	-6.3	-7.3	2	6	1	4
15	-0.4	-6.3	-11.5	-14.0	2	1	7	4
16	-0.2	-6.2	-6.2	-7.9	2	1	7	6
17	-0.3	-6.2	-9.5	-11.6	2	1	5	8
18	-0.2	-5.7	-6.2	-9.4	2	8	1	4
19	-0.3	-6.2	-9.5	-11.7	2	1	7	9
20	-0.1	-6.2	-6.2	-8.5	2	1	9	6
21	-0.2	-6.2	-11.8	-13.4	2	1	10	8
22	-0.1	-5.8	-6.1	-8.7	2	10	1	8
23	-0.2	-6.1	-9.5	-11.8	2	1	7	11
24	-0.1	-6.1	-6.1	-11.5	2	1	11	10
25	-0.2	-6.1	-9.5	-11.9	2	1	9	12
26	-0.1	-5.8	-6.1	-8.9	2	12	1	8
27	-0.1	-6.1	-11.9	-13.6	2	1	13	5
28	-0.1	-6.1	-6.1	-9.0	2	1	13	10
29	-0.1	-6.1	-9.5	-11.9	2	1	9	14
30	-0.1	-5.9	-6.1	-11.1	2	14	1	8
31	-0.1	-6.1	-9.5	-11.9	2	1	11	15
32	-0.1	-6.1	-6.1	-9.1	2	1	15	10
33	-0.1	-6.1	-11.9	-13.8	2	1	16	7
34	0.0	-5.9	-6.1	-9.2	2	16	1	12
35	-0.1	-6.1	-9.5	-11.9	2	1	11	17
36	0.0	-6.1	-6.1	-12.9	2	1	17	14
37	-0.1	-6.1	-9.5	-12.0	2	1	13	18
38	0.0	-5.9	-6.1	-9.3	2	18	1	12
39	-0.1	-6.1	-12.0	-13.8	2	1	19	16
40	0.0	-6.1	-6.1	-9.3	2	1	19	14
41	-0.1	-6.1	-9.5	-12.0	2	1	13	20
42	0.0	-6.0	-6.1	-11.6	2	20	1	10
43	-0.1	-6.1	-9.5	-12.0	2	1	15	21
44	0.0	-6.1	-6.1	-9.3	2	1	21	14
45	0.0	-6.0	-12.0	-16.6	2	1	22	19
46	0.0	-6.0	-6.0	-9.4	2	22	1	16
47	0.0	-6.0	-9.5	-12.0	2	1	15	23
48	0.0	-6.0	-6.0	-13.4	2	1	23	10
49	0.0	-6.0	-9.5	-12.0	2	1	17	24
50	0.0	-6.0	-6.0	-9.4	2	24	1	16
51	0.0	-6.0	-12.0	-13.9	2	1	25	20
52	0.0	-6.0	-6.0	-9.4	2	1	25	18
53	0.0	-6.0	-9.5	-12.0	2	1	17	26
54	0.0	-6.0	-6.0	-11.8	2	26	1	14
55	0.0	-6.0	-9.5	-12.0	2	1	19	27
56	0.0	-6.0	-6.0	-9.4	2	1	27	18
57	0.0	-6.0	-12.0	-13.9	2	1	28	11
58	0.0	-6.0	-6.0	-9.4	2	28	1	20
59	0.0	-6.0	-9.5	-12.0	2	1	19	29
60	0.0	-6.0	-6.0	-16.1	2	1	29	26
61	0.0	-6.0	-9.5	-12.0	2	1	21	30
62	0.0	-6.0	-6.0	-9.4	2	30	1	20
63	0.0	-6.0	-12.0	-13.9	2	1	31	13
64	0.0	-6.0	-6.0	-9.4	2	1	31	22
65	0.0	-6.0	-9.5	-12.0	2	1	21	32
66	0.0	-6.0	-6.0	-11.9	2	32	1	16
67	0.0	-6.0	-9.5	-12.0	2	1	23	33
68	0.0	-6.0	-6.0	-9.5	2	1	33	22
69	0.0	-6.0	-12.0	-13.9	2	1	34	28
70	0.0	-6.0	-6.0	-9.5	2	34	1	24
71	0.0	-6.0	-9.5	-12.0	2	1	23	35
72	0.0	-6.0	-6.0	-13.7	2	1	35	14
73	0.0	-6.0	-9.5	-12.0	2	1	25	36
74	0.0	-6.0	-6.0	-9.5	2	36	1	24
75	0.0	-6.0	-12.0	-16.8	2	1	37	11
76	0.0	-6.0	-6.0	-9.5	2	1	37	26
77	0.0	-6.0	-9.5	-12.0	2	1	25	38
78	0.0	-6.0	-6.0	-11.9	2	38	1	20
79	0.0	-6.0	-9.5	-12.0	2	1	27	39
80	0.0	-6.0	-6.0	-9.5	2	1	39	26
81	0.0	-6.0	-12.0	-13.9	2	1	40	32
82	0.0	-6.0	-6.0	-9.5	2	40	1	28
83	0.0	-6.0	-9.5	-12.0	2	1	27	41
84	0.0	-6.0	-6.0	-13.8	2	1	41	34
85	0.0	-6.0	-9.5	-12.0	2	1	29	42
86	0.0	-6.0	-6.0	-9.5	2	42	1	28
87	0.0	-6.0	-12.0	-13.9	2	1	43	17
88	0.0	-6.0	-6.0	-9.5	2	1	43	30
89	0.0	-6.0	-9.5	-12.0	2	1	29	44
90	0.0	-6.0	-6.0	-11.9	2	44	1	22
91	0.0	-6.0	-9.5	-12.0	2	1	31	45
92	0.0	-6.0	-6.0	-9.5	2	1	45	30
93	0.0	-6.0	-12.0	-14.0	2	1	46	19
94	0.0	-6.0	-6.0	-9.5	2	46	1	32
95	0.0	-6.0	-9.5	-12.0	2	1	31	47
96	0.0	-6.0	-6.0	-13.8	2	1	47	38
97	0.0	-6.0	-9.5	-12.0	2	1	33	48
98	0.0	-6.0	-6.0	-9.5	2	48	1	32
99	0.0	-6.0	-12.0	-14.0	2	1	49	40
100	0.0	-6.0	-6.0	-9.5	2	1	49	34
128	0.0	-6.0	-6.0	-9.5	2	1	63	42
1920	0.0	-6.0	-6.0	-16.9	2	1	959	274
1968	0.0	-6.0	-6.0	-14.0	2	1	983	394

Table 11.3 *Calculated Second Fractional Spur*

Fractional Denominator	In-Band Fractional Spur				Fractional Numerator			
	Worst-case	2nd Worst	3rd Worst	4th Worst	Worst	2nd Worst	3rd Worst	4th Worst
2	-23.0	x	x	x	1	x	x	x
3	x	x	x	x	x	x	1	x
4	-20.0	x	x	x	1	x	x	2
5	-11.3	-15.5	x	x	2	1	x	x
6	-3.9	-13.5	x	x	3	1	x	x
7	-5.3	-10.4	-12.3	x	3	2	1	x
8	-4.0	-11.7	x	x	3	1	x	x
9	-1.6	-11.2	-11.2	-11.2	3	1	4	2
10	-2.5	-10.9	x	x	3	1	x	x
11	-2.1	-7.7	-10.6	-12.2	3	4	1	2
12	-0.9	-10.5	-10.5	x	3	1	5	x
13	-1.5	-7.2	-10.3	-12.2	3	5	1	4
14	-1.3	-10.2	-13.4	x	3	1	5	x
15	-0.6	-4.8	-10.1	-10.1	3	6	1	4
16	-1.0	-10.1	-13.6	-15.0	3	1	7	5
17	-0.8	-6.7	-10.0	-12.1	3	7	1	5
18	-0.4	-9.9	-9.9	-9.9	3	7	1	5
19	-0.7	-6.6	-9.9	-12.1	3	8	1	4
20	-0.6	-9.9	-15.7	-16.6	3	1	9	7
21	-0.3	-5.4	-7.3	-9.8	3	9	6	8
22	-0.5	-9.8	-13.8	-15.9	3	1	5	9
23	-0.5	-6.4	-9.8	-12.1	3	10	1	5
24	-0.2	-7.9	-9.8	-9.8	3	9	1	7
25	-0.4	-6.3	-9.7	-12.1	3	11	1	7
26	-0.4	-9.7	-13.8	-16.2	3	1	11	7
27	-0.2	-5.7	-9.4	-9.7	3	12	6	10
28	-0.3	-9.7	-13.8	-17.9	3	1	5	9
29	-0.3	-6.3	-9.7	-12.1	3	13	1	8
30	-0.1	-8.5	-9.7	-9.7	3	9	11	1
31	-0.3	-6.2	-9.7	-12.1	3	14	1	7
32	-0.2	-9.7	-13.9	-16.5	3	1	7	5
33	-0.1	-5.8	-8.7	-9.7	3	15	12	1
34	-0.2	-9.7	-13.9	-16.5	3	1	13	15
35	-0.2	-6.2	-9.6	-12.1	3	16	1	8
36	-0.1	-9.6	-9.6	-9.6	3	13	1	11
37	-0.2	-6.2	-9.6	-12.1	3	17	1	10
38	-0.2	-9.6	-13.9	-16.6	3	1	7	5
39	-0.1	-5.8	-8.9	-9.6	3	18	12	14
40	-0.2	-9.6	-16.6	-18.5	3	1	11	13
41	-0.1	-6.1	-9.6	-12.1	3	19	1	11
42	-0.1	-9.0	-9.6	-9.6	3	15	1	13
43	-0.1	-6.1	-9.6	-12.1	3	20	1	10
44	-0.1	-9.6	-13.9	-16.7	3	1	17	13
45	-0.1	-5.9	-9.6	-9.6	3	21	16	1
46	-0.1	-9.6	-13.9	-16.7	3	1	19	7
47	-0.1	-6.1	-9.6	-12.1	3	22	1	11
48	-0.1	-9.1	-9.6	-9.6	3	15	17	1
49	-0.1	-6.1	-9.6	-12.1	3	23	1	13
50	-0.1	-9.6	-16.7	-18.7	3	1	21	17
51	0.0	-5.9	-9.2	-9.6	3	24	18	1
52	-0.1	-9.6	-13.9	-16.7	3	1	11	7
53	-0.1	-6.1	-9.6	-12.1	3	25	1	14
54	0.0	-9.6	-9.6	-9.6	3	19	1	17
55	-0.1	-6.1	-9.6	-12.1	3	26	1	13
56	-0.1	-9.6	-13.9	-18.8	3	1	23	19
57	0.0	-5.9	-9.3	-9.6	3	27	18	20
58	-0.1	-9.6	-13.9	-16.8	3	1	11	17
59	-0.1	-6.1	-9.6	-12.0	3	28	1	14
60	0.0	-9.3	-9.6	-9.6	3	21	1	19
61	-0.1	-6.1	-9.6	-12.0	3	29	1	16
62	-0.1	-9.6	-14.0	-16.8	3	1	13	27
63	0.0	-6.0	-9.6	-9.6	3	30	22	1
64	-0.1	-9.6	-14.0	-16.8	3	1	25	27
65	-0.1	-6.1	-9.6	-12.0	3	31	1	17
66	0.0	-9.3	-9.6	-9.6	3	21	23	1
67	-0.1	-6.1	-9.6	-12.0	3	32	1	16
68	-0.1	-9.6	-14.0	-16.8	3	1	13	19
69	0.0	-6.0	-9.4	-9.6	3	33	24	1
70	0.0	-9.6	-18.9	-20.5	3	1	23	13
71	0.0	-6.1	-9.6	-12.0	3	34	1	17
72	0.0	-9.6	-9.6	-9.6	3	25	1	23
73	0.0	-6.1	-9.6	-12.0	3	35	1	19
74	0.0	-9.6	-14.0	-16.8	3	1	29	11
75	0.0	-6.0	-9.4	-9.6	3	36	24	26
76	0.0	-9.6	-14.0	-16.8	3	1	31	33
77	0.0	-6.1	-9.6	-12.0	3	37	1	20
78	0.0	-9.4	-9.6	-9.6	3	27	1	25
79	0.0	-6.1	-9.6	-12.0	3	38	1	19
80	0.0	-9.6	-16.8	-18.9	3	1	11	27
81	0.0	-6.0	-9.6	-9.6	3	39	28	1
82	0.0	-9.6	-14.0	-16.8	3	1	17	23
83	0.0	-6.0	-9.6	-12.0	3	40	1	20
84	0.0	-9.4	-9.6	-9.6	3	27	29	1
85	0.0	-6.0	-9.6	-12.0	3	41	1	22
86	0.0	-9.6	-14.0	-16.8	3	1	35	25
87	0.0	-6.0	-9.4	-9.6	3	42	30	1
88	0.0	-9.6	-14.0	-16.8	3	1	17	13
89	0.0	-6.0	-9.6	-12.0	3	43	1	23
90	0.0	-9.6	-9.6	-9.6	3	31	1	29
91	0.0	-6.0	-9.6	-12.0	3	44	1	22
92	0.0	-9.6	-14.0	-16.8	3	1	19	39
93	0.0	-6.0	-9.4	-9.6	3	45	30	32
94	0.0	-9.6	-14.0	-16.9	3	1	37	13
95	0.0	-6.0	-9.6	-12.0	3	46	1	23
96	0.0	-9.4	-9.6	-9.6	3	33	1	31
97	0.0	-6.0	-9.6	-12.0	3	47	1	25
98	0.0	-9.6	-14.0	-19.0	3	1	19	33
99	0.0	-6.0	-9.6	-9.6	3	48	34	1
100	0.0	-9.6	-16.9	-19.0	3	1	29	33
128	0.0	-9.6	-14.0	-16.9	3	1	25	37
1920	0.0	-9.5	-9.5	-9.5	3	639	641	1
1968	0.0	-9.5	-9.5	-9.5	3	657	1	655

Table 11.4 *Calculated Third Fractional Spur*

Measurement of Fractional Spurs Explanation

The spur table on the next page shows the values for *InBandSpur* calculated by subtracting the rolloff from the measured uncompensated fractional spurs. Note that experiments were done on other PLLs to verify that these trends concerning uncompensated fractional spurs roughly apply to compensated fractional spurs. The table illustrates the following properties concerning fractional spurs.

The darker shaded boxes indicate that no fractional spur is present. This rule can be generalized by stating that a fractional spur is present whenever the greatest common divisor of the fractional numerator and fractional denominator divide evenly into the spur order. For instance, for a fraction of 3/15, the greatest common divisor is 3. Because 3 does not divide evenly into 1 or 2, the first and second fractional spurs will not be present. However, the third fractional spur will be present.

Without loss of generality, the fraction can be reduced to lowest terms. For instance, the first fractional spur when the fraction is 2/5 is roughly the same as the second fractional spur when the fraction is 4/10 and these spurs would occur at the same offset frequency from the carrier.

The worst-case for the first fractional spur is when the fractional numerator is one or one less than the fractional denominator. For the second fractional spur, it is two or two less than the fractional spur. For the third fractional spur, it is three or three less than the fractional denominator. These worst-case spurs are shown in bold text with heavier outlines for the boxes.

InBandSpur$_{Uncompensated}$ (Worst-case) = 1.6 dBc. The exception, which is unlikely to occur in a practical situation, is when the fraction is ½ or can be reduced to ½. In this case, the number is closer to 4.

If the fractional denominator is prime, if one can avoid the worst-case spur, then the next worse case spur is about 6 dB better. There are applications, such as CDMA, where the frequency planning makes this possible. If the denominator is not prime, then even more improvement is possible. These next worse case spurs are in the unshaded boxes. The lightly shaded boxes indicate that a fractional spur is present, but it is neither worst-case nor next to worse case. Note that it becomes more difficult to determine which spurs are next to worse case for higher order spurs. The next worse case fractional spur tends to be about 6 dB lower than the main fractional spur and occurs when the fractional numerator is about ½ of the fractional denominator. However, as the fractional denominator increases, this seems to be closer to $1/3^{rd}$ of the fractional denominator.

			Fractional Denominator														
			2	3	4	5	6	7	8	9	10	11	12	13	14	15	16
First Fractional Spur	Fractional Numerator	1	3.1	2.4	2.0	1.4	1.3	1.4	1.6	1.6	1.5	1.7	1.6	2.0	1.8	1.7	1.8
		2		2.0		-2.7		-5.6		-7.4		-9.3		-10.7		-12.4	
		3			2.0	-2.6		-3.8	-6.1		-7.0	-8.6		-8.8	-10.4		-11.1
		4				1.4		-3.8		-4.0		-6.9		-7.0		-9.1	
		5					1.6	-5.8	-6.2	-4.0		-4.1	-10.1	-10.2	-7.3		-7.3
		6						1.5				-3.9		-4.2			
		7							1.4	-7.5	-6.8	-6.9	-9.9	-3.8		-4.6	-12.7
		8								1.5		-8.6		-10.2		-4.7	
		9									1.5	-9.3		-7.2	-7.5		-12.2
		10										1.6		-9.0			
		11											1.8	-10.7	-11.2	-10.1	-7.8
		12												1.7			
		13													1.0	-11.9	-10.7
		14														1.3	
		15															1.3
Second Fractional Spur	Fractional Numerator	1	1.4	-1.1	-2.7	-3.8	-4.2	-4.7	-4.9	-5.2	-5.1	-5.4	-5.3	-5.7	-5.4	-5.4	
		2	1.1	4.8	1.5	2.7	1.4	1.8	1.2	1.6	1.4	1.4	1.4	1.3	1.1	1.6	
		3		-1.1	1.4		-5.9	-4.5		-8.4	-7.7		-10.4	-10.2		-12.1	
		4			-2.5	2.5		-6.1		-7.6	-2.6	-9.7		-10.9	-5.6	-12.3	
		5				-3.2	1.1	-4.0	-8.3		-8.6	-4.7	-7.4	-11.8		-11.9	
		6					-3.2	1.8		-2.6	-9.4		-8.9	-3.6		-6.1	
		7						-3.7	1.2	-8.7	-9.4	-4.4	-9.6	-52.2	-9.1	-4.6	
		8							-3.6	1.4	-7.2		-7.9	-3.5	-10.1		
		9								-3.6	1.1		-11.4	-11.5		-4.4	
		10									-3.6	1.4	-10.4	-5.5		-6.3	
		11										-3.6	1.2	-9.5	-12.4	-12.4	
		12											-3.4	1.4			
		13												-3.5	1.4	-12.1	
		14													-3.2	1.6	
		15														-3.4	
Third Fractional Spur	Fractional Numerator	1		1.9	-2.7	-5.6	-6.3	-7.1	-7.8	-7.8	-8.2	-8.9	-9.2	-9.1	-9.6	-9.3	
		2			1.1		-4.7		-7.3		-9.9		-11.8		-13.7		
		3		1.4	1.6	4.6	1.5	1.2	2.5	1.2	1.1	1.8	1.3	1.1	1.6	0.4	
		4			-2.7	1.0		-6.8		-5.0		-9.5		-8.8			
		5				-4.4	-3.5	1.0	-6.6		-9.7	-7.5	-5.0	-11.9		-12.8	
		6					-5.7		2.5		-10.0		-10.5		-2.8		
		7						-6.1	-6.3	1.1	-4.2	-8.2	-11.5		-12.1	-11.8	
		8							-6.5		0.8		-4.5		-13.1		
		9								-6.6	-8.7	1.6	-9.3	-11.6	-2.6	-12.1	
		10									-6.9		1.2				
		11										-6.9	-10.0	0.0	-8.1	-13.6	
		12											-6.7		1.6		
		13												-7.7	-11.4	0.9	
		14													-7.4		
		15														-7.4	

Table 11.5 *Measured In-Band Fractional Spurs*

The Fractional Modulus Game

One may first think that the worst-case fractional spur is the one that should be considered. In general, this is usually true. However, there are cases where it is possible to avoid using a fractional modulus of one due to good frequency planning. For PLLs used in the CDMA standard, this is often the case. The fractional modulus game becomes possible when the fractional denominator used exceeds the number of channels the PLL synthesizer has to tune to. In CDMA, a fractional denominator of 1968 is often used, even there are only 23 channels. From the table, avoiding a fractional numerator of one or one less than the fractional denominator yields about a 6 dB improvement in fractional spurs. Note that the next worse case usually occurs when the fractional numerator is one-half or one-third of the fractional denominator. Figure 11.1 shows fractional spurs measured with the LMX2364 fractional N PLL with the compensation turned off and with a fractional denominator of 100. Note that 1 and 99 are the worst case for this first fractional spur, and if these two numerators can be avoided, then approximately a 10 dB benefit can be realized. Using the fractional spur tables or the fractional spur chart predict that this would yield a benefit of 9.5 dB. The next worst case are 33 and 67, as predicted by simulations.

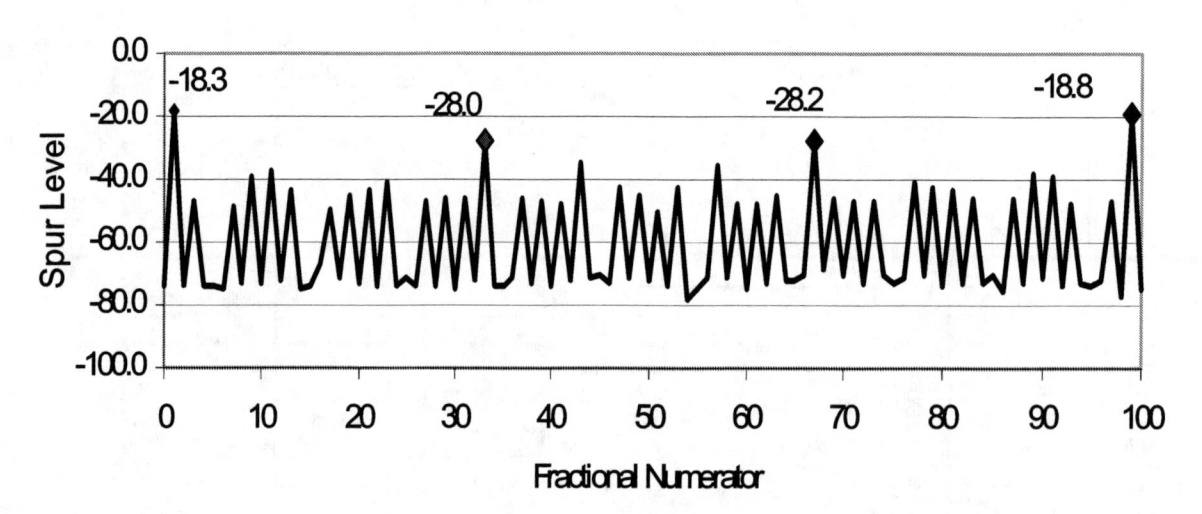

Figure 11.1 *Uncompensated Fractional Spurs as a Function of the Fractional Numerator*

Uncompensated Fractional Spur Model

Comparing the mathematical model with the measured results for fractional spurs, there is fairly good agreement, especially with the relative levels of the spurs. However, they do differ from by a mysterious factor of 1.6 dB. Therefore, the fractional spur model will be based around the mathematical model with an added factor of 1.6 dB. Uncompensated fractional spurs can be predicted as follows:

$$Spur_{Fractional\ Uncompensated} = InBandSpur_{Uncompensated} + 1.6\ dB + rolloff \qquad (11.5)$$

Note that the number used for the *InBandSpur* in this model is the theoretical number. The mysterious factor of 1.6 dB is to account for the difference between the theoretical model and actual measured data.

Impact of Fractional Compensation on Fractional Spurs

Fractional compensation can be very complicated with many exceptions. It can be impacted by the PLL voltage or by which prescaler is used. There may be different modes and settings to further complicate the matter. Higher order delta sigma modulation can be viewed as a digital means to fractional compensation as well. On some PLLs, it is possible to disable the fractional compensation so that its impact can be clarified. On other PLLs, it cannot be disabled. Uncompensated fractional spurs tend to be more predictable and relatively constant over the VCO tuning voltage.

Compensated Fractional Spurs

Until now, the discussion has been focused on uncompensated fractional spurs because the impact of fractional compensation can vary based on the type of fractional compensation. One complexity that compensation tends to add is that it makes spurs vary as a function of tuning voltage and output frequency. For some parts, expressing the fraction different ways, such as 4/64 as opposed to 1/16 can make a difference in fractional spur levels. After dealing with compensated fractional spurs, one will probably come to the realization that predicting fractional spurs to the last dB without actually testing the circuit is folly. In general, models concerning fractional spurs can be within a few dB on a good day, but there are certainly exceptions where the models can be far more off than that. Some of the relative relationships regarding fractional spurs tend to hold better. Despite all of these limitations, prediction of fractional spurs is still worthwhile, but should always be tested against measured data. Although this sort of simplifies the model, compensation usually reduces uncompensated spur level by some fixed amount. It therefore makes sense to quantify compensated fractional spurs by their in-band spur performance. Also, it is usually practical to measure this quantity directly.

$$\textit{Fractional Spur} = \textit{InBandSpur} + \textit{rolloff} \qquad (11.6)$$

PLL	*InBandSpur*	Comments
Uncompensated Fractional PLL	+1.6	Based on measurements from the LMX2364 with compensation disabled. This does not apply to fractional spur levels worse than about −12 dBc.
LMX2350/52/53/54	− 15	InBandSpur for second fractional spur is closer to −12 dBc
LMX2364	−18	InBandSpur for the second fractional spur is closer to −13 dBc. Spur level is sensitive to fractional denominator.
LMX2470	−20 to −50 −35 typical	The fractional spurs on this part are better when the comparison frequency is around 20 MHz and a fractional denominator greater than 100. There is benefit expressing fractions with higher fractional denominators, even if the mathematical values are equivalent. Outside the loop bandwidth this number steadily increases

Figure 11.2 *In-Band Compensated Fractional Spurs for Various PLLs*

Delta Sigma Fractional Spurs

The fractional spur model derived so far works quite good for fractional parts. For non-delta sigma PLLs, subtracting a fixed value for compensation seems to work quite well. However, for delta sigma fractional PLLs, there are many exceptions. For one thing, the value for the in-band spur can vary a little with the loop bandwidth, even though the spur of interest is well within the loop bandwidth in all cases. Another factor is that the spurs do not always roll off at as fast at a rate as the closed loop response predicts. For instance, the LMX2470 delta sigma PLL fractional spurs follow the closed loop transfer function very closely, but after about twice the loop bandwidth, they roll off at a rate of 20 dB/decade, as opposed to the 40 dB/decade that the closed loop transfer function predicts. Other delta sigma parts also exhibit this same characteristic, although the offset frequency for which the spurs track the closed loop bandwidth is different. Within the loop bandwidth or close to the loop bandwidth the models work better, but outside the loop bandwidth they tend to be optimistic because spurs for delta sigma parts often, but not always roll off at a rate that is less than what the loop filter would roll off the spurs.

Delta Sigma Sub-Fractional Spurs

For delta sigma PLLs, it is possible to get spurs at offsets less than the channel spacing. In general, the higher order the fractional modulator is, the worse these spurs will be. They often occur at one-half and/or one-fourth of where a traditional fractional spur would be. Although these spurs may look unattractive on a spectrum analyzer, they may not be as damaging to performance as traditional fractional spurs because they do not occur at an offset equal to the channel spacing; this is application specific. The bigger concern with these sub-fractional spurs may often times be their impact on RMS phase error. Depending on the standard and the level of these spurs, they can have a more dominant impact on RMS phase error than phase noise in many cases. The levels of sub-fractional spurs are difficult to predict and have many application-specific dependencies. Among the factors that may influence these spurs are TCXO power level, output frequency, tuning voltage, delta sigma modulator order, various selectable bits for the PLL, and the loop filter.

Filtering and the order of the delta sigma modulator are two factors that usually have a large impact on these spurs. Because the delta sigma modulator pushes lower frequency spur energy out to higher frequencies, it is important to have filtering at these higher frequencies in order to prevent these higher frequency products from being translated back down to lower frequencies. In general, the order of the PLL loop filter should be one greater than the order of the delta sigma modulator. However, this is just a rule of thumb. If the loop bandwidth is too wide, then higher order filters may not be able to eliminate these spurs. Also, if the loop bandwidth is sufficiently narrow, then a loop filter of lower order may be sufficient.

Conclusion

The art of predicting fractional spurs is not an exact science. They tend to be more exceptions and variations than with integer spurs. One property of fractional spurs that tends to hold up is that they are virtually immune to leakage currents. Delta sigma PLLs can add more layers of complexity by introducing spurs at sub-multiples of where traditional fractional spurs would be.

Because it is difficult dealing with fractional spurs, the concepts were first developed by exploring uncompensated fractional spurs. Unlike their compensated counterparts, uncompensated spurs are very regular and easy to predict. Once these models are understood, compensation is simply treated as reducing the uncompensated spur by some fixed amount.

Some may wonder why it is worth any effort trying to predict fractional spurs at all. There is definitely value in this effort, but there is also no substitute for bench measurements. The models presented here are intended as tools.

Chapter 12   On Non-Reference Spurs and their Causes

Introduction

Much has been said about reference spurs and fractional spurs. This chapter investigates other types of spurs and their causes. The value of doing this is so that when a spur is seen, its causes and fixes can be investigated. Although many types of spurs are listed, most of these spurs are not usually present. Since a lot of these spurs occur in dual PLLs, the main PLL will always refer to the side of a dual PLL on which the spur is being observed, and the auxiliary PLL will refer to the side of a dual PLL that is not being observed. This chapter discusses general good tips for dealing with spurs, and then goes into categorizing the most common types, their causes, and their cures.

Tips for Good Decoupling and Good Layout

To deal with board-related cross talk, there are several steps that can be taken. Be sure to visit wireless.national.com and download the evaluation board instructions to see typical board layouts. In addition to this, there are the following additional suggestions:

Good Decoupling: By this it is meant to have several capacitors on both the *Vcc* and charge pump supply lines. The charge pump supply lines are the most vulnerable to noisy signals. Place a 100 pF, 0.01 µF, and a 0.1 µF capacitor on each of these lines to deal with noise at a wide range of frequencies. It may seem that these capacitances simply add in parallel to form a 0.111 µF capacitor, but in fact, they are all necessary since the larger capacitors have more problems responding to high frequency signals and may have a higher ESR. In general, it is important to have the smaller capacitors closest to the PLL chip to deal with high frequency noise, but it is fine and often more convenient to place the larger capacitors farther away. The trace between the smaller capacitor and the larger capacitor adds inductance, which is good. It is also a good idea to put a series resistor to help deal with low frequency noise. This resistor should be chosen so that the voltage drop across it is around 0.1 V. 18 Ω is a typical value, but this should be dependent on the current that flows through this resistor.

Good Layout: Be sure to keep the charge pump supply lines and the VCO tuning voltage lines away from noisy signals. Try to make ground vias close to the part and try to minimize the sharing of ground vias. Placing a ground plane in the board to separate the top and bottom layer also can help reduce cross talk effects.

Good Loop Filter Design: Higher order loop filters and filters with narrower loop bandwidth are more effective in reducing spurs of all sorts – not just reference spurs.

Cross Talk vs. Non-Cross Talk Related Spurs

For the purposes of this discussion, the spurs will be divided into two categories. Cross talk related spurs refer to any spur that is caused by some source other than the PLL that finds its way to VCO output. Non-cross talk related spurs refer to spurs that are caused by some inherent behavior in the PLL. The first step in diagnosing a spur is to determine whether or not it is a cross talk related spur. The way that this is done is by eliminating all potential causes of the cross talk spur and checking if the spur goes away.

Cross Talk Related Spurs

In general, signals that are either low frequency, or close to the PLL output frequency are the most likely to cause this type of spurs. Whenever two sinusoidal signals enter a non-linear device an output signal at the sum and the difference of these frequencies will be produced. This result can be derived by writing the first three general terms for the Taylor series and observing that the square term gives rise to these sum and difference frequencies. It therefore follows that frequencies that are low in frequency, or frequencies that are close to the PLL output frequency are the ones that cause the most problems with cross talk related spurs. Several different types of the cross talk related spur will now be discussed

Auxiliary PLL Cross Talk Spur

Description: This spur only occurs in dual PLLs and is seen at a frequency spacing from the carrier equal to the difference of the frequencies of the main and auxiliary PLL (or sometimes a higher harmonic of the auxiliary PLL). This spur is most likely to occur if the main and auxiliary sides of a dual PLL are close in frequency. If the auxiliary PLL is powered down, but the auxiliary VCO is running, then this spur can dance around the spectrum as the auxiliary frequency VCO drifts around.

Cause: Parasitic capacitances on the board can allow high frequency signals to travel from one trace on the board to another. This happens most for higher frequencies and longer traces. There could also be cross talk within the chip. The charge pump supply pins are vulnerable to high frequency noise.

Diagnosis: One of the best ways to diagnose this spur is to tune the auxiliary side of the PLL while observing the main side. If the spur moves around, that is a good indication that the spur being observed is of this type. Once this type of spur is diagnosed, then it needs to be determined if the spur is related to cross talk on the board, or cross talk in the PLL. Most PLLs have a power down function that allow one to power down the auxiliary side of a PLL, while keeping the main side running. If the auxiliary side of the PLL is powered down, and the spur reduces in size substantially, this indicates cross talk in the PLL chip. If the spur stays about the same magnitude, then this indicates that there is cross talk in the board.

Cure: Read the section on how to deal with board related cross talk.

Crystal Reference Cross Talk Spur

Description: This spur is visible at an offset from the carrier equal to some multiple of the crystal reference frequency. Often times, there is a whole family of spurs that often occur at harmonics of the crystal reference frequency. In this case, the odd harmonics are usually stronger than the even harmonics.

Cause: This spur can be caused by excessive gain of the inverter in the crystal oscillator. Sometimes, this inverter is integrated unto the PLL chip. When any oscillator has excessive gain, it can give rise to harmonics. The reason that the odd harmonics are often stronger is that the oscillator often produces a square wave or a clipped sine wave, which has stronger odd harmonics. Figure 12.1 shows the structure of a typical crystal oscillator. Note that ***Lm*** (motional inductance), ***Cm*** (motional capacitance), and ***Cp*** (parallel capacitance) represent the circuit equivalent of a quartz crystal.

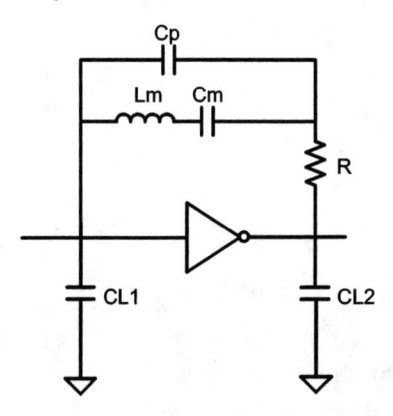

Figure 12.1 *A Typical Crystal Oscillator Circuit*

Diagnosis: The best way to diagnose this spur is to use a signal generator in place of the crystal. If spur level is impacted, then this is an indication that the oscillator inverter has excessive gain. Note that on some of National Semiconductor's PLLs, the inverting buffer is included on the PLL chip, while on others, it is not. If the power level to the chip is reduced, then this decreases the gain of the buffer, which theoretically should decrease the level of this type of spur.

Cure: In addition to the suggestions about good decoupling and layout, there are several things that may reduce these spur levels

1. *Decrease the gain of the inverting buffer*

This may sound sort of ridiculous at first, but if the part is run at a lower VCC power supply voltage, then the gain of the inverter is decreased. Also, some of National Semiconductor's PLLs, such as the LMX160x family have only a single inverter stage as opposed to a triple inverter stage.

2. Supply an external inverter

Using a separate inverter for the crystal, or using the inverter from some other component, such as the microprocessor could also be a fix.

3. Increase the value of the Resistor, R

In the above diagram, increasing the value of **R** can account a little bit for the excessive inverter gain. If **R** is increased too much, the circuit simply will not oscillate. Note that in many inverter circuits $R = 0 \; \Omega$.

4. Try unequal load capacitors

Usually, the load capacitors, CL1, and CL2 are chosen to be equal, but in this case it might improve the spur level to make CL2 > CL1. This is because the output of the inverter is a square wave, so anything to round out the edges can help.

5. Layout and filtering

Be sure to read the layout tips and also consider filtering the noisy signal on the board.

External Cross Talk Spur

Description: This spur appears and is unrelated to the auxiliary PLL output. Often times, when the main PLL is tuned to different frequencies, this spur moves around.

Cause: This type of spur is caused by some frequency source external to the PLL. Common external sources that can cause these spurs are: computer monitors (commonly causes spurs at the screen refresh rate of 30 – 50 kHz), phones of all sorts, other components on the board, florescent lights, power supply (commonly causes spurs in multiples of 60 Hz), and computers. Long signal traces can act as an antenna and agitate this type of spur.

Diagnosis: To diagnose this spur, start isolating the PLL from all potential external noise sources. Switch power supplies. Turn off computer monitors. Go to a screen room. Disconnect the auxiliary VCO and power down the auxiliary PLL. By trial and error, external noise sources can be ruled out, one by one.

Cure: To eliminate this spur, remove or isolate the PLL from the noise source. As usual, these spurs are layout dependent, so be sure to read the section on good layout. Also consider using RF fences to isolate the PLL from potential noise sources.

Non-Cross talk Related Spurs

These spurs are caused by something other than cross talk on the board. Some common examples are discussed below:

Greatest Common Multiple Spur

Description: This spur occurs in a dual PLL at the greatest common multiple of the two comparison frequencies. For example, if one side was running with a 25 kHz comparison frequency, and the other side was running with a 30 kHz comparison frequency, then this spur would appear at 5 kHz. In some cases, this spur can be larger on certain output frequencies.

Cause: The reason that this spur occurs is that the greatest common multiple of the two comparison frequencies corresponds to the event that both charge pumps come on at the same time. This result can be derived by considering the periods of the two comparison frequencies. When both charge pumps come on, they produce noise, especially at the charge pump supply pins, which gives birth to this spur.

Diagnosis: A couple telltale signs of this type of spur is it is always spaced the same distance from the carrier, regardless of output frequency. However, keeping the output frequency the same, but changing the comparison frequency causes this spur to move around. Just be sure that when changing the comparison frequencies for diagnostic purposes, you are also changing the greatest common multiple of the two comparison frequencies.

Cure: This spur can be treated effectively by putting more capacitors on the Vcc and charge pump supply lines. Be sure that there is good layout and decoupling around these pins. Also consider changing the comparison frequency of the auxiliary PLL.

Phantom Reference Spur

Description: The phantom reference spur is characterized by a ghastly increase in the reference spurs right after switching frequencies. After the frequency is changed, it takes an excessively long time for the reference spurs to settle down. This spur is more common at lower comparison frequencies.

Cause: Some of this can be possibly explained by deceptive measurements from the equipment, such as using the video averaging function on a spectrum analyzer. It can also be caused by leaky capacitors in the loop filter. Other theories suggest that it is related to undesired effects from the loop filter capacitors, such as dielectric absorption.

Diagnosis: This can be observed on a spectrum analyzer. Just be very careful that it is not some sort of averaging effect of the spectrum analyzer. The output of the spectrum analyzer is power vs. frequency, which is really intended to be a still time sort of measurement. It may be helpful to test the equipment measuring some other spur to make sure that this is really the PLL and not the equipment.

Cure: Designing with higher quality capacitors helps a lot. In particular, the capacitor *C2* tends to be the culprit for causing this spur. Common capacitor types listed in order of improving dielectric properties are: tantalum, X7R, NP0, and polypropeline. Also, using a fractional N PLL can possibly help, since the fractional spurs tend to be less impacted by the effects of charge pump leakage and non-ideal capacitor dielectrics.

Prescaler Miscounting Spur

Description: This spur typically occurs at half the comparison frequency. It can also occur at one-third, two-thirds, or some fractional multiple of the comparison frequency. It can have mysterious attributes, such only occurring on odd channels.

Cause: This spur is caused by the prescaler miscounting. Things that cause the prescaler to miscount include poor matching to the high frequency input pin, violation of sensitivity specifications for the PLL, and VCO harmonics. Be very aware that although it may seem that the sensitivity requirement for the PLL is being met, poor matching can still agitate sensitivity problems and VCO harmonic problems. Note also that there is an upper sensitivity limitation on the part.

To understand why the prescaler miscounting causes spurs, consider fractional N averaging. Since the prescaler is skipping counts on some occasions and not skipping counts on another, it produces spurs similar to fractional spurs.

Diagnosis: Since miscounting ties in one way or another to sensitivity, try varying the voltage and/or temperature conditions for the PLL. Since sensitivity is dependent on these parameters, any dependency to supply voltage or temperature point to prescaler miscounting as the cause of the spur. Changing the *N* counter between even and odd values can also sometimes have an impact on this type of spur caused by the *N* counter miscounting, and can be used as a diagnostic tool.

Also be aware that *R* counter sensitivity problems can cause this spur as well. One way to diagnose *R* counter miscounting is to change the *R* counter value just slightly. If the spur seems sensitive to this, then this may be the cause. If a signal generator is connected to the reference input, and the spur mysteriously disappears, then this suggests that the *R* counter miscounting is the cause of the spur.

Cure: To cure this problem, it is necessary to fix whatever problem is causing the prescaler to miscount. The first thing to check is that the power level is within the specifications of the part. After that, consider the input impedance of the PLL. For many PLLs, this tends to be capacitive. Putting an inductor to match the imaginary part of the PLL input impedance at the operating frequency can usually fix impedance matching issues. Be also aware of the sensitivity and matching to the VCO harmonics, since they can also cause a miscount. Try to keep the VCO harmonics –20 dBm or lower in order to reduce the chance of the PLL miscounting the VCO harmonic.

Prescaler Oscillation Spur

Description: This spur typically occurs far away from the carrier at an offset frequency of approximately the output frequency divided by the prescaler value. In most applications, it is not a concern because it is out of band.

Cause: This spur is caused internally by the output frequency being divided by the prescaler. It comes out through the high frequency input pin.

Diagnosis: This spur is sensitive to isolation between the VCO and the PLL. The frequency offset is a good indicator of this spur. Be sure to power down the PLL and make sure the spur goes away to verify it is not a cross talk issue.

Cure: If this spur is a problem, the solution is in providing greater isolation for the high frequency input pin of the PLL. The most basic way is to put a pad with sufficient attenuation. The issue with this is that the attenuation of the pad may be limited by the sensitivity limits of the PLL. Another approach is to put an amplifier, which increases isolation. Yet a third approach is to use a directional coupler, but this is frequency specific and costs layout area.

VCO Harmonic Spurs

Description: This spur occurs at multiples of the output frequency. All VCOs put out harmonics of some kind. This spur can cause problems if there is very poor matching to the high frequency input of the PLL. Note also in some cases, the higher harmonic can have better matching and sensitivity performance than the fundamental. This can cause mysterious noisy behaviors. In general, it is good to have the second harmonic 20 dB down if possible, but that is very dependent on the matching and the sensitivity of the PLL.

Cause: VCOs are part specific in what level of harmonics they produce, but they all produce undesired harmonics of the fundamental frequency.

Diagnosis: These spurs appear at the VCO frequency and multiples thereof. Change the VCO frequency, and see if the spurs still appear at multiples of the VCO output.

Cure: If the VCO harmonics cause a problem there are several things that can be done to reduce their impact. They can be low pass filtered with LC or RC filters. A resistor or inductor can be placed in series at the fin pin to prevent them from causing the prescaler to miscount. Just make sure that there is good matching and that the spur level at the fin pin is as low as possible. Note also that the many PLLs do not have a 50 Ω input impedance. Treating it as such often creates big problems with the VCO harmonics.

Conclusion

In this chapter some, but not all causes of spurs have been investigated. Although it is difficult to predict the levels of non-reference spurs, their diagnosis and treatment is what is really matters. Non-reference spurs tend to be a thing that requires a lot of hands on type of diagnostics, and process of elimination is sometimes the only way to figure out what is the real cause.

Chapter 13   PLL Phase Noise Modeling and Behavior

Introduction

This chapter investigates the causes and behaviors of phase noise. The goal of this chapter is to develop a mathematic model in order to understand phase noise. The model is developed by taking factors that influence phase noise one factor at a time. The factors considered, in the order they are presented are: N divider, comparison frequency, charge pump gain, VCO noise, 1/f noise, resistor noise, and other noise sources. The more of these factors that are accounted for, the more accurate the model, but also the more work to calculate the phase noise. The first step is to study the PLL noise before it is shaped by the loop filter.

Accounting for the N Divider and Closed Loop Transfer Function – Phase Noise Floor

The most basic model for phase noise assumes that it is some constant that is multiplied by the closed loop transfer function. In order to make things even simpler, the closed loop transfer function can be approximated by the N divider value, provided that the offset frequency is within the loop bandwidth. Since phase noise is caused by a noise voltage, the noise power would be proportional to N^2, hence this implies that phase noise varies as *20•log(N)*. If the phase noise is not within the loop bandwidth, then the closed loop transfer function cannot be approximated in this way. There is nothing wrong with this theory, however, it disregards the effects of the phase detector. Phase noise floor is this constant value and is calculated below.

Phase Noise (approximation for in-band) = PhaseNoiseFloor + 20•log|N| (13.1)

Phase noise (everywhere) = PhaseNoiseFloor + 20•log|CL(f)|

This phase noise model works for older phase detectors, such as the mixer phase detector, but for the modern charge pump PLL, it does not account for the variation in the comparison frequency, which is too relevant of a factor to ignore.

1 Hz Normalized Phase Noise Floor (PN1Hz)

Unlike the phase noise floor model, the model taking into account the comparison frequency is accurate enough be practical in many situations. Assuming a digital 3-state phase-frequency detector, this will put out more noise at higher comparison frequencies. The phase-frequency noise also tends to be the dominant noise source, which is proportional to the comparison frequency. However, the comparison frequency is inversely proportional to N.   So therefore the noise due to the phase detector degrades in accordance with *10•log(Fcomp)*.   Therefore, phase noise can be predicted using the 1 Hz Normalized Phase Noise (*PN1Hz*), which is part-specific and assumes the highest charge pump current.

(13.2)

Phase Noise (in-band) = PN1Hz + 20•log|N| + 10•log|Fcomp|
Phase noise (everywhere) = PN1Hz + 20•log|CL(f)| + 10•log|Fcomp|

Prediction of Close in Phase Noise as a Function of N for a Fixed Output Frequency

What the above formula implies is that if the output frequency is fixed, but the *N* counter value is changed, then both the *N* counter value and the comparison frequency are changing. This could be the case when one is using a fractional PLL and wanting to know the impact of changing the *N* counter, which corresponds to raising **Fcomp**. The phase noise in this case varies as **10•log(N)**. For instance, if the comparison frequency is doubled, which corresponds to making the *N* counter half, and the output frequency is held constant, then the phase noise would improve by 3 dB, not 6 dB.

Modeling PN1Hz for Variable Charge Pump Gains – KφKnee

In the closed loop transfer function for the charge pump noise, there is a factor of *1/Kφ*. The implication of this factor is that the phase noise is better for higher charge pump gains. In general, this is true, however, because the charge pump noise also increases with the gain, the true relationship is not as the closed loop transfer function predicts. In general, the easiest way to model this is to try the different charge pump gains. In general, one should use the highest charge pump gain unless there is a compelling reason not to. Some of the reasons not to use the highest charge pump gain could be that the loop filter capacitors get too large to be practical, or the highest charge pump gain needs to be reserved for Fastlock. Experiments show that increasing the charge pump gain helps to a point, but then there are diminishing returns. If the gain is too low, then doubling the charge pump gain gives a 3 dB improvement, but if the gain is high enough, then there is not much more benefit to raising the comparison frequency. **KphiKnee** is a term that describes the knee of this charge pump current vs. phase noise trade-off. For instance, if one is operating with a charge pump current equal to **KphiKnee**, then the theoretical phase noise benefit of increasing the charge pump current from this level to infinite is 3 dB. In summary, the 1 Hz Normalized phase noise is not constant over charge pump current and can be modeled what it theoretically would be for an infinite charge pump gain with a correction term for the charge pump gain.

$$PN1Hz(K\phi) = PN1Hz + 20\bullet\log |1 + KphiKnee/K\phi| \qquad (13.3)$$

Adding in the 1/f Noise into the PLL Noise Estimate

So far, it has been assumed that the phase noise is perfectly flat within the loop bandwidth, even to offsets as low as 1 Hz. In actuality, it is more correct to say that the phase noise is flat past a certain offset, but for frequencies less than some offset, it degrades at a rate of 10 dB/decade. For many applications, it may be that the offset frequency for which this 1/f noise comes into play is low enough that it does not impact the system performance. In applications where this is relevant, one needs to make sure that the measurement equipment is capable of making such a measurement. Assuming that this phase noise close to the carrier is relevant and it can be measured, the next challenge is that it varies more than other phase noise. For instance, if the crystal reference level is increased, then this noise can decrease. It could also be sensitive to voltage. Using a higher *R* divider value can often reduce this noise. One also has to be very careful in assessing this 1/f noise. In many cases, it can be the noise from the crystal reference source, especially if a signal generator is being used.

In general, the 1/f noise is most severe when operating at higher comparison frequencies or when using fractional parts. One rule of thumb is that if the phase noise offset is 0.5% of the comparison frequency, the effects of this 1/f noise can be ignored. For a fractional part operating at 20 MHz comparison frequency, this number works out to be 100 kHz, which is quite far from the carrier. In general, this noise decreases by 10 dB/decade and a simple way to characterize this is to normalize it to 10 kHz offset frequency and a 1 GHz PLL output frequency. To add this noise to the traditional flat noise, one converts both to non-dB units, adds them, and then converts back to dB. Note that the *N* counter value does not multiply this noise.

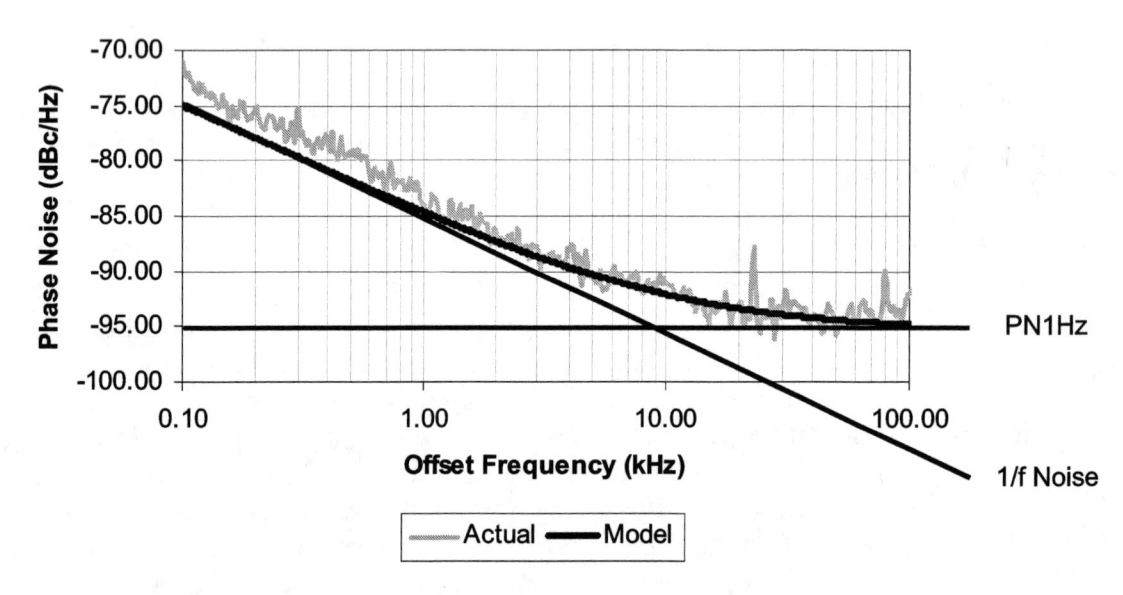

Figure 13.1 *Unshaped Phase Noise Example with the LMX2470*

Figure 13.1 shows the phase noise of the National Semiconductor LMX2470 delta sigma PLL measured at 2.44 GHz output frequency and a 5 MHz comparison frequency. The loop bandwidth for this measurement was around 500 kHz, so this phase noise is all in-band. Note that the 1/f noise is added to the 1 Hz Normalized PLL Noise AFTER it is shaped by the closed loop transfer function.

*Another Interpretation of 1/f Noise -- **FcompKnee***

If one considers a fixed offset frequency and PLL output frequency, but varies the comparison frequency, the phase noise seems to improve with the comparison to a point, and then reach some sort of saturation. Using the 10 kHz Normalized phase noise model for 1/f noise would predict this. So one way that this could happen is that it is simply another way to interpret the PN10kHz Number. For this case, the **FcompKnee** frequency can be calculated from the PN10kHz number from the following relationship.

$$PN10kHz + 10 \cdot \log\left|\frac{10kHz}{Offset}\right| = PN1Hz + 10 \cdot \log|FcompKnee| + 20 \cdot \log\left|\frac{1GHz}{FcompKnee}\right| \quad (13.4)$$

However, the 1/f phase noise model sometimes gives a deceptive results. Some parts have very flat phase noise, even down to about 10 Hz. For these parts, the 1 Hz Normalized phase noise degrades for higher comparison frequencies. In this case, the 1/f noise model may not be the best choice and the **FcompKnee** model may be a better fit. The **FcompKnee** is the comparison frequency for which the 1 Hz normalized phase noise is degraded by 3 dB. Furthermore, increasing the comparison frequency from this point to infinity only results in a 3 dB improvement in the overall phase noise. In other words:

$$PN1Hz(Fcomp) = PN1Hz + 10 \bullet \log\left|1 + \frac{Fcomp}{FcompKnee}\right| \qquad (13.5)$$

More Corrections for 1/f noise -- Noise Plateau

If the modeling of the 1/f noise was not already complicated enough, now it is time to start making exceptions to this noise. For many parts, the 1/f noise model is adequate down to offset frequencies of 100 Hz, but for others, it gives a false impression for lower offset frequencies. In general, for frequency offsets of less than 1 kHz, the behavior is part-specific. For instance, for the LMX2330 family of PLLs, the 1/f noise stops at about 1 kHz offset and then no longer degrades for offset frequencies down to 10 Hz. The phase noise is flat from offset frequencies on the order of 1 kHz to 10 Hz from the carrier. One way to model this is with a noise plateau. What this means is that for phase noise offsets less than the noise plateau frequency, the noise is flat.

Simplified Definition of 1 Hz Normalized Phase Noise

The easiest way to model the PLL noise is with a simple constant, the 1 Hz Normalized Phase Noise floor (***PN1Hz***). This is independent of offset. The next step is to add in a term for the 1/f noise, and then finally, the noise plateau number can be used. There is a separate chapter with a phase noise example using these concepts. For the sake of simplicity, it is much easier to quantify this noise metric of a PLL with a single number. By default, this will assume a 2 kHz offset frequency, a 200 kHz comparison frequency, and maximum charge pump current. Now this is also called the 1 Hz Normalized Phase Noise Floor, so one needs to keep track of if this number accounts for the charge pump current or not.

Accounting for the VCO Noise

The VCO noise has a slope that increases as the offset to the carrier is decreased. However, at these close offsets, the PLL noise is usually dominant. It therefore makes matters much easier to deal with if the VCO noise is modeled as a fixed value at 10 kHz, called VCO10kHz and with a 20db/decade slope. Recall that the VCO has a high pass transfer function.

Accounting for the Resistor Noise and OP AMP Noise (if applicable)

This noise source tends to present the most problem near the loop bandwidth and at higher offset frequencies. If the charge pump gain is increased, this noise source is decreased. This noise source is covered in depth in the appendix.

Accounting for the Crystal Reference Noise

For most applications, it is assumed that this noise source is not dominant. However, it can definitely show up in some applications. The most common situation is when one tries to drive the PLL with a signal generator, which has much higher noise than a TCXO. If a signal generator is used to drive the PLL, one trick is to increase the R counter and the signal generator frequency, such that this noise is divided down more.

Phase Noise Constants for Various National PLLs

The phase noise performance is part specific. Table 13.1 contains typical phase noise data for various National Semiconductor PLLs. It is true that the dividers, Crystal Reference, and VCO contribute to the in-band phase noise, but these are typically dominated by the noise of the phase detector. Since the phase detector noise is dependent on the comparison frequency, this table is normalized for what the phase detector noise would theoretically be for a 1 Hz comparison frequency (***PN1Hz***). This table is based on sample data taken from evaluation boards using an automated test program that varies the comparison frequency and charge pump currents.

Table 13.1 gives a rough indication of how one PLL will perform against another. The expected dB difference is simply the difference in the numbers from the table. Note for the fractional N PLLs (LMX2350/2352/2354/2364/LMX2470), the phase noise floor can be deceptive. Since the fractional N capability allows one to use a higher reference frequency, the actual phase noise tends to be better, despite the fact that the phase noise floor is degraded. This is because the value of N will be smaller. So one should be cautious about comparing phase noise of fractional parts without also considering the benefit of the fractional modulus.

For this table, another term, PN1HzNorm, is introduced. This is what the phase noise would be at a 2 kHz offset operating at the maximum charge pump current. This number is useful to use if one does not want to deal with the corrections for 1/f noise and charge pump gain.

Part	Side	PN1Hz dBc/Hz	PN10kHz (1GHz Norm) dBc/Hz	KphiKnee μA	KphiMax μA	PN1HzNorm Offset=2kHz Fcomp=200kHz Kphi=KphiMax dBc/Hz	Plateau kHz	Comments
LMX1600/01/02	RF/IF	-209.3	-99.3	130	1600	-208.5	1.0	
LMX2323/4	RF	-215	-97.7	0	4000	-214.5	1.0	
LMX2306/16/26	RF	-214	-104.8	1000	1000	-210.9	1.0	
LMX2310/11/12	RF	-213	-93.5	1400	4000	-211.1	1.0	
LMX1501/11	RF	-208.0	-86.1	0	4000	-207.0	1.0	
LMX2301/05/15/20/25	RF	-208.0	-86.1	0	4000	-207.0	1.0	
LMX2330/31/32 (A and L)	RF/IF	-214.0	-93.8	1000	4000	-212.4	1.0	
LMX2330/31/32U	RF/IF	-214.8	-94.6	1000	4000	-213.2	1.0	Model is ~2 dB worse than actual at 1 mA and *Fcomp*>2 MHz
LMX2346/47	RF	-217.8	-99.6	0	4000	-217.1	1.0	
LMX2430/31/32	RF	-217.8	-99.6	0	4000	-217.1	-	
LMX2350/52/53	RF	-203.5	-89.3	70	1600	-203.0	-	
LMX2350/52/54	IF	-210.0	-98.0	300	800	-208.6	-	
LMX2354	RF	-207.0	-88.0	50	1600	-206.1	-	Disabling fractional compensation improves phase noise about 7 dB.
LMX2364	RF	-208.8	-95.8	800	16000	-208.4	-	Disabling fractional compensation improves phase noise about 7 dB.
LMX2364	IF	-214.0	-108.9	500	4000	-213.3	-	
LMX2470	RF	-215.0	-104.0	2000	1600	-211.4	-	Note that phase noise is operating below the current knee.
LMX2470	IF	-210.0	-100.1	0	4000	-209.3	-	

Table 13.1 *1 Hz Normalized Phase Noise Floor for Various National Semiconductor PLLs*

Sample Calculation

The first example shows the phase noise of the LMX2350 PLL using a VARIL1960U VCO. For this example, because the comparison frequency was low and the charge pump gain was the maximum setting, the phase noise is modeled by –203.0 dBc/Hz plus any shaping due to the closed loop transfer function. This model shows how the PLL, VCO, and resistor noise interact. Note that it is necessary to subtract off *10•Log(Resolution Bandwidth)* from the plots of the spectrum analyzer to get the phase noise.

For the second example, the LMX2470 was used. The loop bandwidth was very wide, about 300 kHz. The comparison frequency was 5 MHz, and the charge pump gain was 1600 μA. The output frequency was 2.44 GHz. This shows the 1/f noise behavior. For this part, the 1/f noise is much more observable due to the high comparison frequency and low phase noise. For this example, the resolution bandwidth and spectrum analyzer correction factors are already figured into the measured data.

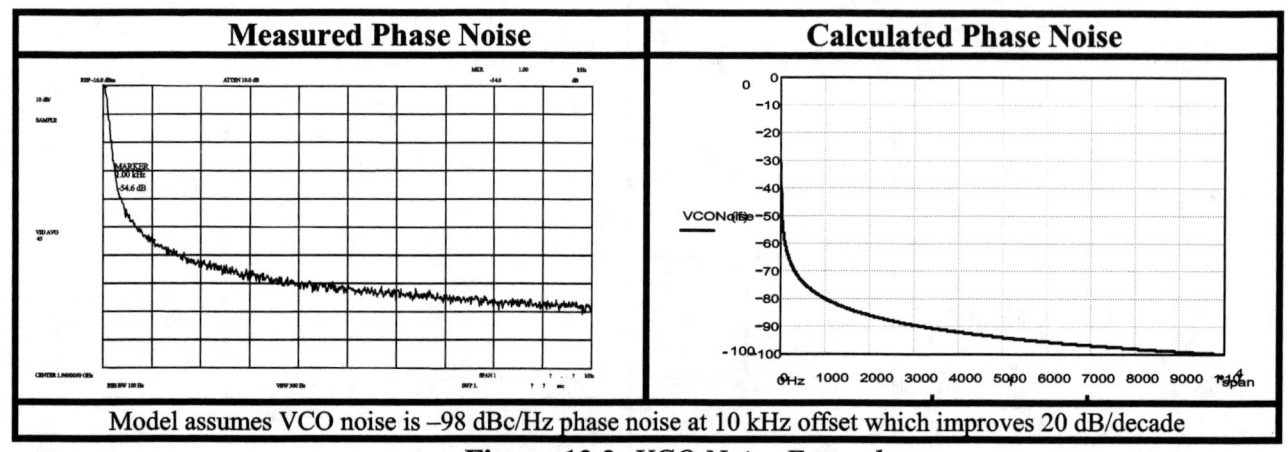

Model assumes VCO noise is –98 dBc/Hz phase noise at 10 kHz offset which improves 20 dB/decade

Figure 13.2 *VCO Noise Example*

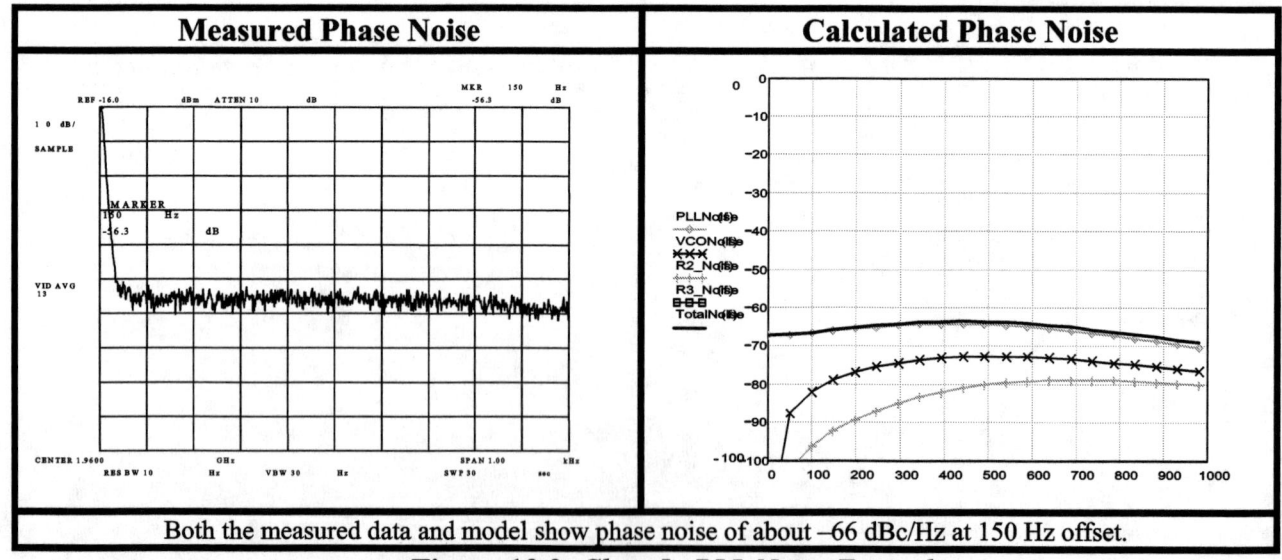

Both the measured data and model show phase noise of about –66 dBc/Hz at 150 Hz offset.

Figure 13.3 *Close In PLL Noise Example*

Measured Phase Noise	Calculated Phase Noise

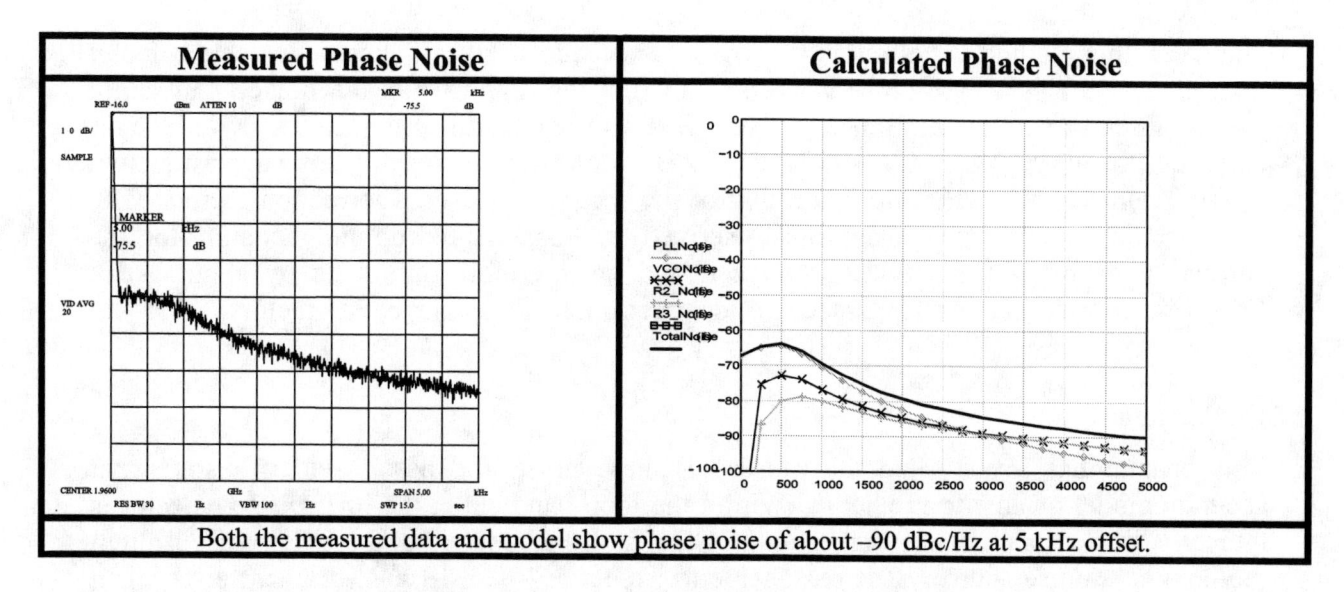

Both the measured data and model show phase noise of about −90 dBc/Hz at 5 kHz offset.

Figure 13.4 *Far Out Phase Noise Example*

More Issues with Phase Noise Modeling and Measurement

Crosstalk in Dual PLLs

In the dual PLL, it has been found that the optimal phase noise performance is when the other side of the PLL is unused, powered down, and with no VCO connected. If this is the case, then this results in a 2 dB improvement from what the table predicts. The table assumes that the other PLL is powered down, but the VCO is connected. If the other side is powered up and running, then the degradation in phase noise may be a dB or two worse than the table predicts.

Issues with Input Sensitivity

There are many ways to cause the phase noise to be worse than predicted. One possible cause of this is when either the VCO or crystal power levels are insufficient to drive the counters. For the high frequency VCO, matching problems can also cause an input sensitivity problem. These phase noise numbers assume that the VCO and crystal power levels are sufficient to drive the counters, and that there are no problems matching the VCO to the prescaler input pin. Although rare, there are also PLLs for which the input buffer contributes phase noise and for these PLLs, a higher crystal oscillator drive level is required for optimal performance.

Spectrum Analyzer Correction Factors

A common way of measuring phase noise using a spectrum analyzer is as follows:

$$\text{Phase Noise} = \text{Carrier Power} - \text{Noise Power} - 10 \bullet Log(\text{Resolution Bandwidth}) \qquad (13.6)$$

However, this method is not entirely correct. Spectrum analyzers have a correction factor that is added to the phase noise to account for the log amplifier in the device and minor errors caused due to the difference between the noise bandwidth and 3 dB bandwidth. This correction factor is in the order of about 2.5 dB. Many spectrum analyzers have a function called "Mark Noise", which does account for the spectrum analyzer correction factors. The part-specific numbers for phase noise derived in this chapter do not account for the correction factor of the spectrum analyzer, and are therefore optimistic by about 2 dB. Numbers reported in this chapter account for spectrum analyzer correction factors.

Conclusion

This chapter has investigated the causes of phase noise and has provided a somewhat accurate model of how to predict it. Within the loop bandwidth, the PLL phase detector is typically the dominant noise source, and outside the loop bandwidth, the VCO noise is often the dominant noise source. It is reasonable to expect a +- 0.5 dB measurement error when measuring phase noise. Phase noise can vary from board to board and part to part, but typically this variation is in the order of a few dB.

References

Lascari, Lance   *Accurate Phase Noise Prediction in PLL Frequency Synthesizers*
       Applied Microwave & Wireless Vol.12 No. 5. May 2000

Lascari, Lance   *Mathcad PLL Phase Noise Simulation Tool,*
       http://home.rodchester.rr.com/lascari/lancepll.zip

Phase noise Measurement of PLL Frequency Synthesizers National Semiconductor
       Application Note 1052

Appendix A: Phase Noise for Resistors and Active Devices

Noise Voltages

Resistors and active devices such as op-amps generate noise voltages. In the case of an op-amp, the noise voltage should be specified. In the case of a resistor, this noise voltage is the thermal noise generated by the resistor. Recall that the thermal noise generated by a resistor is:

$$R\_Noise = \sqrt{4 \bullet T_0 \bullet K \bullet R} \qquad (13.7)$$

T_0 = *Ambient Temperature in Kelvin = 300 Kelvin (typically)*

K = *Boltzman's Constant = 1.380658 X 10^{-23} (Joule/Kelvin)*

R = *Resistor Value in Ohms*

Note that in both the case of the resistor and op-amp, the units are $\frac{V}{\sqrt{Hz}}$. Since phase noise is normalized to a 1 Hz bandwidth, one can disregard the denominator and consider the units to be in Volts.

Transfer Function for the Noise Voltage

Once the noise voltage is known, an open-loop transfer function, T(s), can be written which relates this noise voltage to the voltage it would generate for an open loop system at the VCO tuning line. To account for the closed loop system, one can simply divide this by the open loop transfer function of the VCO (Lascari 2000). In deriving the transfer function, T(s), it is simplifies calculations if one remembers that all the grounds are connected and draws a short between them. In the case of a resistor noise transfer function, the resistor noise can be considered to be acting on either side of the resistor. The actual transfer functions will not be derived here, since the formulas are shown in the design example at the end of this chapter.

Translating the Noise Voltage to a dBc/Hz number for Phase Noise

This explanation is found in reference listed by Lance Lascari. In a similar way that leakage-based reference spur was shown to relate to the modulation index of the signal, the modulation index is applied here to derive the phase noise. **Vnoise** represents the noise voltage that would be generated at the VCO input for an open loop system, f is the frequency, and G is the open loop transfer function. Note that it is necessary to multiply the noise voltage by a factor of $\sqrt{2}$, since these are expressed as RMS, and not Peak to Peak.

$$\text{Phase Noise} \approx 20 \bullet \log\left(\frac{\beta}{2}\right) \qquad \text{For } \beta << 1 \qquad (13.8)$$

$$\beta = \frac{\sqrt{2} \bullet Vnoise \bullet Kvco}{f} \bullet \frac{|T(2\bullet\pi\bullet i\bullet f)|}{\left|1 + \frac{G(2\bullet\pi\bullet i\bullet f)}{N}\right|}$$

Resistor noise becomes a problem when the resistors in the loop filter get too large. The resistor noise tends to have the greatest contribution at frequencies close to the loop bandwidth. It can also have some contribution outside the loop bandwidth. Using a higher current gain or Fractional *N* PLL can reduce the impact of resistor noise. Op-amp noise can also add considerable phase noise, especially if the op-amp is not very low noise.

Chapter 14   RMS Phase Error and Derived Noise Quantities

Introduction

This chapter discusses RMS Phase error, how to calculate it, the relevance it has in digital communications, and how to minimize it. It also discusses the signal to noise ratio of a PLL and it's relationship to phase noise.

What is RMS Phase Error?

There are three different ways of visualizing RMS phase error. It can be visualized in the time domain, in the frequency domain, or in a constellation diagram. These different interpretations of RMS phase error are all related and discussed below.

RMS Phase Error In the Time Domain

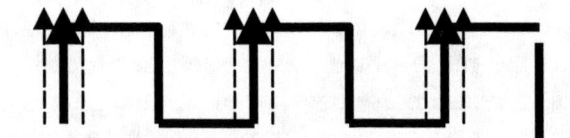

Figure 14.1 *Illustration of RMS Phase Error of a Signal in the Time Domain*

Figure 14.1 shows a square wave Note that the rising edges of the square wave do not always occur at exactly the time they should, but have a random phase error that can be either positive or negative. Now the average value of this phase error is zero, but the standard deviation is nonzero and is called the RMS phase error. Recall for the normal distribution, approximately 68% of the area of the normal distribution curve is within one standard deviation of the mean. This means that if one was to take a random sample of the starting phase, 68 % of the time it would be within the RMS phase error. Notice how the rising edges of the signal do not always start at the time they should, but sort of jitters. For this reason, RMS phase error and phase noise are often referred to as "phase jitter". Although the output of a PLL tends to be a sine wave (instead of a square wave), there is little loss of generality here, because the sine wave goes through counters that turn it into a square wave.

For an example, consider a 10 MHz signal with 5 degrees RMS phase error. Since the period of this signal is 0.1 µS, a 5 degree RMS phase error imply a normally distributed random phase shift which has a standard deviation of 1.339 nS. A term sometimes used to describe this phase shift is *jitter*.

RMS Phase Error Calculation from Frequency Domain

Formula for Relating Spectral Density to RMS Phase Error

RMS Noise is calculated by integrating the phase noise, taking the square root, and then converting this number from radians to degrees. The limit, *a*, tends to be very close to the carrier, and the limit, *b*, tends to be much farther away, typically outside the loop bandwidth. Assuming *b* to be infinite gives a reasonable approximation to RMS phase error. The RMS phase error in degrees is calculated as:

$$RMS\ Phase\ Error = \frac{180}{\pi} \cdot \sqrt{2 \cdot \int_a^b L(f) \cdot df} \qquad (14.1)$$

Derivation of RMS Phase Error Formula

The derivation for formula (14.1) will now be given. Recall that phase noise is measured in dBc/Hz on a spectrum analyzer, which shows the output power vs. frequency. Since phase noise is measured at a particular frequency output, it can be thought of as the ratio of the carrier frequency power to the noise power, expressed in a decibel scale. Actually, it is more correct to view this as a phase noise density, even though it is commonly just referred to as phase noise. To obtain the total phase error, the phase noise (density) is integrated over the whole frequency spectrum. The factor of two is there to account for the phase noise on both sides of the carrier.

Since the spectrum analyzer displays power vs. frequency, and not voltage vs. frequency, it is necessary to take the square root of the integrated product to obtain an RMS (Root Mean Square) error. Recall that in statistics, the standard deviation of a continuous random variable is obtained by integrating the square of the probability distribution function and then applying the square root. An analogous procedure is used in the calculation of RMS phase error and this is why the RMS phase error relates to the standard deviation of the phase error. Since the number obtained is a dimensionless value in radians, it is necessary to convert this to degrees.

Approximate RMS Phase Error Calculation

To calculate the RMS noise correctly, the spectral density needs to be known. However, it is sometimes convenient to use a rule of thumb to simplify calculations, or in situations where the VCO noise is unknown. One good rule of thumb is to assume that the phase noise decreases 20 dB/decade from the PLL loop bandwidth. This approximation is shown in Figure 14.2 .

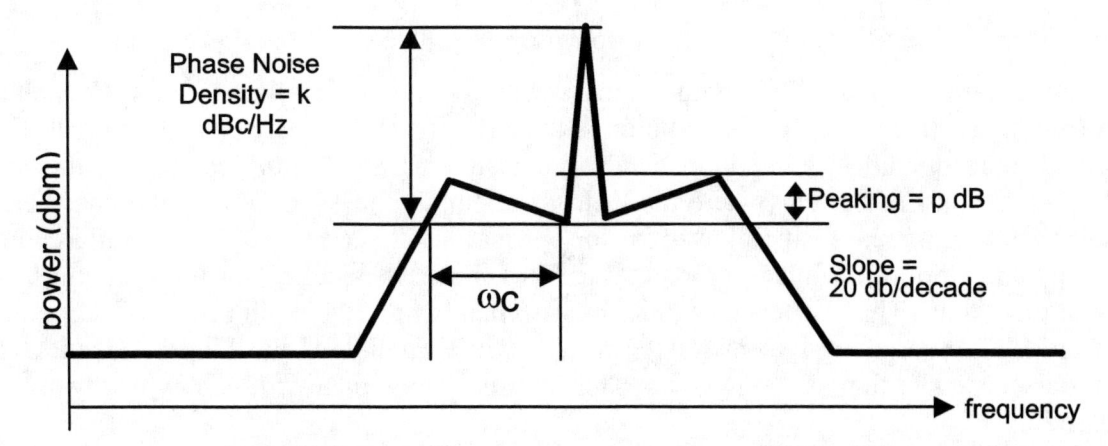

Figure 14.2 *Typical Phase Noise Spectral Plot for a PLL*

Approximate Calculation of RMS Phase Error

To calculate the RMS Phase Error, formula (14.1) will now be applied. Since the phase noise density, **k**, is expressed in dBc/Hz, it is necessary to convert this from decibels to scalar units before the integration is performed.

$$RMSnoise = \frac{180}{\pi}\sqrt{2 \cdot \left(\int_0^{fc} 10^{k/10} \cdot \left(1 + \left[10^{p/10} - 1\right] \cdot \frac{f}{fc}\right) \cdot df + \int_{fc}^{\infty} 10^{(k+p)/10 - 20 \cdot \frac{\log(f - fc + 1)}{10}} \cdot df\right)} \quad (14.2)$$

$$= \frac{180}{\pi} \cdot 10^{k/20} \sqrt{fc \cdot \left(1 + 10^{p/10}\right) + 10^{p/10} \cdot 2}$$

Note that for the purposes of simplifying calculations, it was assumed that the phase noise peaks at the loop bandwidth, but in actuality, the peaking occurs slightly before the loop bandwidth as shown in Figure 14.2. This causes these estimations to be slightly lower than they actually should be. Three dB is a typical value for the peaking, which would be typical for 45 degrees of phase margin, however, it makes sense to use a value slightly higher, since this will help compensate for the fact that the estimates are slightly low. A good value of peaking to use is 4 dB. For the sake of simplicity, it makes sense to introduce approximations. Note that the second term under the square root is very small compared to the first term for any loop bandwidth that is reasonable. If one neglects the second term and assumes 4 dB of peaking, the formula can be greatly simplified.

$$RMSnoise = 107 \cdot 10^{k/20} \sqrt{fc} \quad (14.3)$$

If 0 dB of peaking, then multiply this result by 75%. If 3 dB of peaking is assumed, then multiply this result by 92%. If 10 dB of peaking is assumed, multiply this result by 177%.

For a System with 10 kHz loop bandwidth, and –80 dBc/Hz phase noise, (assume 4dB peaking):

$$RMSPhaseError = 107 \cdot 10^{-80/20} \cdot \sqrt{10000} = 1.1° \quad (14.4)$$

Choice of Loop Bandwidth for Optimal RMS Phase Error

This formula implies that a PLL with a narrower loop bandwidth will have less RMS phase error, but in fact this is only true to a point. The validity of the approximations used in the above formula degrades if the loop bandwidth is too narrow. After decreasing the loop bandwidth beyond this point (where the PLL noise equals the VCO noise), the phase error actually starts to increase. It follows that for optimal RMS phase error, one should choose the loop bandwidth of the system such that the PLL noise is equal to the free-running VCO noise at that point. This is because within the loop bandwidth, the main noise contribution is from the PLL (everything except for the VCO), while outside of the loop bandwidth, the main noise contribution is the VCO. This optimal loop bandwidth is typically a few kilohertz.

Impact of Spurs on RMS Phase Error

In most cases, spurs are outside the loop bandwidth and have only a very small impact on the RMS phase error. However, many fractional PLLs and especially delta sigma PLLs have spurs that can occur inside the loop bandwidth. The way to treat a spur is to assume that all the energy is inside a 1 Hz bandwidth. The impact of a spur depends on the bandwidth. If one considers the phase noise to be flat within the integration bandwidth, then the spur relates to the phase noise in a 10•log(Bandwidth) sense. For instance, a spur that is 40 dB above the noise floor has the same noise energy as the phase noise itself if the phase noise is flat and the integration bandwidth is 10 kHz. Two spurs that are 37 dB above this noise floor would also have equivalent noise energy as the phase noise itself.

RMS Phase Error Interpretation in the Constellation Diagram

If one visualizes the RMS error in the time domain, then it can be seen why this may be relevant in clock recovery applications, or any application where the rising (or falling) edges of the signal need to occur in a predictable fashion. The impact of RMS phase error is more obvious when considering a constellation diagram.

The constellation diagram shows the relative phases of the I (in phase) and Q (in quadrature – 90 degrees phase shift) signals. The I and Q axes are considered to be orthogonal, since their inner product is zero. In other words, for any signal received, the I and Q component can be recovered. Each point on the constellation diagram corresponds to a different symbol, which could represent multiple bits. As the number of symbols is increased, the bandwidth efficiency theoretically increases, but the system also becomes more susceptible to noise. Quadrature Phase Shift Keying (QPSK) is a modulation scheme sometimes used in cellular phones. Figure 14.3 shows the constellation diagram for QPSK.

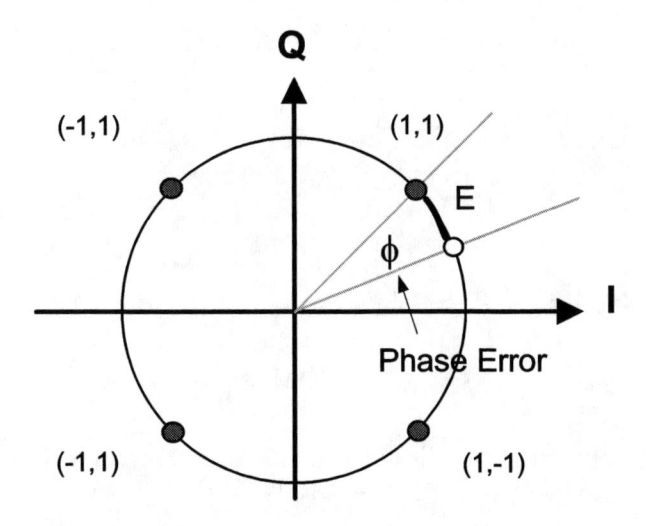

Figure 14.3 *Impact of RMS phase Error Seen on a Constellation Diagram*

Consider an ideal system in which the only noise-producing component is the PLL in the receiver. In this example, the symbol corresponding to the bits (1,1) is the intended message indicated by the darkened circle. However, because the PLL has a non-zero RMS phase error contribution, the received signal is actually the non-filled circle. If this experiment was repeated, then it the result would be that the phase error between the received and intended signal was normally distributed with a standard deviation equal to the RMS phase error. If the RMS phase error of the system becomes too large, it could actually cause a the message to be misinterpreted as (-1, 1) or (1,-1). This constellation diagram interpretation of RMS phase error shows why higher order modulation schemes are more subject to the RMS phase error of the PLL. A real communications system will have a noisy channel and other noisy components, which reduce the amount of RMS phase error of the PLL that can be tolerated.

Other Interpretations of RMS Phase Error

Eye Diagram

One popular way of viewing RMS phase error is the eye diagram. The impact of the RMS phase error on the eye diagram is that it causes it to close up. This means that the decision region is smaller and it is more likely to make an error in which bits were sent.

Error Vector Magnitude (EVM)

Error Vector Magnitude is the magnitude of the vector formed from the intended message and the actual message received (refer to Figure 14.3). This is commonly expressed as a percentage of the error vector relative to the vector formed between the origin and intended message. Referring to Figure 14.3 , assuming the circle has radius R, and applying the law of cosines yields the magnitude of the error vector , E, to be:

$$E = 2 \cdot R^2 - 2 \cdot R^2 \cdot cos(\phi) \qquad (14.5)$$

Assuming that ϕ is small, and using the Taylor series expansion $cos(\phi) = 1 - \phi^2/2$, yields the following relationship between RMS phase error and EVM:

$$EVM \approx 100\% \cdot \left(\frac{\pi}{180}\right) \cdot (RMS\ Phase\ Error\ in\ Degrees) \qquad (14.6)$$

Jitter

Jitter is the time domain interpretation of RMS phase error. It can be calculated by converting the RMS phase error to VCO periods:

$$Jitter = \frac{1}{Fout} \cdot \frac{RMS\ Phase\ Error\ in\ Degrees}{360°} \qquad (14.7)$$

RMS Frequency Error

RMS frequency error is the standard deviation of the frequency error and is calculated as follows:

$$RMS\_Frequency\_Error = \sqrt{2 \cdot \int_a^b L(f) \cdot f^2 df} \qquad (14.8)$$

Signal to Noise Ratio (SNR)

The signal to noise ratio of a PLL refers to the carrier power to the noise power. Since phase noise is expressed in terms of dBc/Hz. Without loss of generality, the signal can be considered to be concentrated in a 1 Hz bandwidth with relative power level of 0 dBc/Hz. The total power for this signal is 0 dBc.

The noise power can be found by integrating the power spectral density, except for the carrier. The lower integration limit, *a*, can be assumed to be 0 Hz, just as long as the carrier is disregarded. The upper integration limit, *b*, is the bandwidth of interest, perhaps the channel spacing. Choosing *b* to be infinite typically serves as a reasonable approximation for general purpose discussions. The signal to noise ratio in dB is therefore:

$$SNR = \frac{1}{2 \cdot \int_0^b L(f) \cdot df} \qquad (14.9)$$

There are other ways to define the signal to noise ratio of a PLL as well. To correctly figure how the SNR of a PLL impacts the SNR of a system is actually a detailed calculation. However, a simple analogy can be used to give a rough idea of how the PLL SNR impacts the system.

Consider an input signal to a mixer:

$$S1 = Si + Ni \qquad (14.10)$$

Where S1 is the total input signal, Si is the desired input signal, and Ni is the undesired input noise. Now assume that the PLL signal is:

$$S2 = Spll + Npll \qquad (14.11)$$

The output signal is therefore the product of the two signals S1 and S2

$$Sout = Spll \bullet Si + Spll \bullet Ni + Si \bullet Npll + Npll \bullet Ni \qquad (14.12)$$

Now the first term is the desired signal power and the last term is negligible. The output signal to noise ratio can therefore be approximated as:

$$SNR = \frac{Spll \bullet Si}{Spll \bullet Ni + Si \bullet Npll} = \frac{SNR1 \bullet SNR2}{SNR1 + SNR2} \qquad (14.13)$$

In the above equation, **SNR1** and **SNR2** represent the signal to noise ratios of **S1** and **S2**, respectively. In an analogous way that two resistances combine in parallel, the lower signal to noise ratio dominates. The above calculations contain some very gross approximations, but they do show how the signal to noise ratio of the PLL can degrade the signal to noise ratio of the whole system.

Conclusion

This chapter has covered various parameters that are derived from the phase noise of the PLL, including RMS phase error. Unlike the phase noise discussed in the previous chapter, the RMS phase error is very dependent on the loop bandwidth of the PLL. RMS phase error is often a parameter of concern in digital communication systems, especially those using phase modulation.

Chapter 15 Transient Response of PLL Frequency Synthesizers

Introduction

This chapter considers the frequency response of a PLL when the N divider is changed. In addition to giving a fourth order model of this event, whose only approximation is the continuous time approximation for the phase detector, it also gives derivations for natural frequency and damping factor, which are used in a second order approximation. It further relates them to loop bandwidth and phase margin. This chapter is intended to give a rigorous mathematical foundation for the transient response of PLL synthesizers. In doing so, it provides a universal model which can be used in place of all of the various rules of thumb, since rules of thumb only work under certain conditions or for certain applications.

Derivation of Transfer Functions

The filter coefficients $A0$, $A1$, $A2$, and $A3$ were discussed in a previous chapter. Recall that the transfer function of the loop filter is as follows:

$$Z(s) = \frac{1+s \cdot T2}{s \cdot \left[A3 \cdot s^3 + A2 \cdot s^2 + A1 \cdot s + A0\right]} \tag{15.1}$$

This leads to the following closed-loop transfer function:

$$CL(s) = \frac{K \cdot N \cdot (1+s \cdot T2)}{s^5 \cdot A3 + s^4 \cdot A2 + s^3 \cdot A1 + s^2 \cdot A0 + s \cdot K \cdot T2 + K} \tag{15.2}$$

$$K = \frac{K\phi \cdot Kvco}{N}$$

It should be noted that the N value to use for the equation is the N value corresponding to the final frequency value, not the starting frequency value or the N value the loop filter was optimized for.

Second Order Approximation to Transient Response

Derivation of Equations

To this point, no approximations have been made, and this form works up to a fourth order loop filter. In this section, $CL(s)$ will be approximated by a second order expression, in order to derive results that give an intuitive feel of the transient response.

It is assumed that these higher order terms are small relative to the lower order terms. The Initial Value Theorem (15.3) suggests that the consequences of ignoring these terms are more on the initial characteristics, such as overshoot, and less on long time behavior, such as lock time. In addition, the impact of neglecting the zero, $T2$, has a noticeable impact on overshoot, but only a minimal impact on lock time.

$$\lim_{s \to \infty} s \bullet Y(s) = \lim_{t \to 0} y(t) \qquad (15.3)$$

The simplified second order expression is:

$$CL(s) \approx \frac{K/A0}{s^2 + s \bullet \left(\frac{K \bullet T2}{A0}\right) + \frac{K}{A0}} \qquad (15.4)$$

Defining

$$\omega n = \sqrt{\frac{K\phi \bullet Kvco}{N \bullet A0}} \qquad (15.5)$$

$$\zeta = \frac{T2}{2} \bullet \omega n \qquad (15.6)$$

It can be seen that the poles of this function are at:

$$-\zeta \bullet \omega n \pm j \bullet \omega n \bullet \sqrt{1-\zeta^2} \qquad (15.7)$$

Now consider a PLL, which is initially locked at frequency *f1*, and then the *N* counter is changed such to cause the PLL to switch to frequency *f2*. It should be noted that the value for *N* that is used in all of these equations should be the value of *N* corresponding to *f2*. This event is equivalent to changing the reference frequency from *f1/N* to *f2/N*. Using inverse Laplace transforms it follows that the frequency response is:

$$F(t) = f2 + (f1 - f2) \bullet e^{-\zeta \bullet \omega n \bullet t} \bullet \left[\cos\left(\omega n \sqrt{1-\zeta^2} \bullet t\right) + \frac{\zeta - R2 \bullet C2 \bullet \omega n}{\sqrt{1-\zeta^2}} \bullet \sin\left(\omega n \sqrt{1-\zeta^2} \bullet t\right) \right] \qquad (15.8)$$

Since the term in brackets has a maximum value of:

$$\frac{1 - 2 \bullet R2 \bullet C2 \bullet \zeta \bullet \omega n + R2^2 \bullet C2^2 \bullet \omega n^2}{\sqrt{1-\zeta^2}} \qquad (15.9)$$

It follows that the lock time in seconds is given by:

$$\text{Lock Time} = \frac{-\ln\left(\frac{tol}{f2-f1} \cdot \frac{\sqrt{1-\zeta^2}}{1 - 2 \cdot R2 \cdot C2 \cdot \zeta \cdot \omega n + R2^2 \cdot C2^2 \cdot \omega n^2}\right)}{\zeta \cdot \omega n} \quad (15.10)$$

Many times, this is approximated by:

$$\text{Lock Time} = \frac{-\ln\left(\frac{tol}{f2-f1} \cdot \sqrt{1-\zeta^2}\right)}{\zeta \cdot \omega n} \quad (15.11)$$

The peak time can be calculated by taking the derivative of expression (15.8) and setting this equal to zero. In this case, calculations are simplified if one assumes that *C2* is sufficiently small. Note that the solution for t=0 is ignored.

$$\text{Peak Time} = \frac{\pi}{\omega n \sqrt{1-\zeta^2}} \quad (15.12)$$

Combining this with (15.8) and (15.9) yields the following:

$$\text{Overshoot} = \frac{f2-f1}{\sqrt{1-\zeta^2}} \cdot e^{-\zeta \cdot \pi / \sqrt{1-\zeta^2}} \quad (15.13)$$

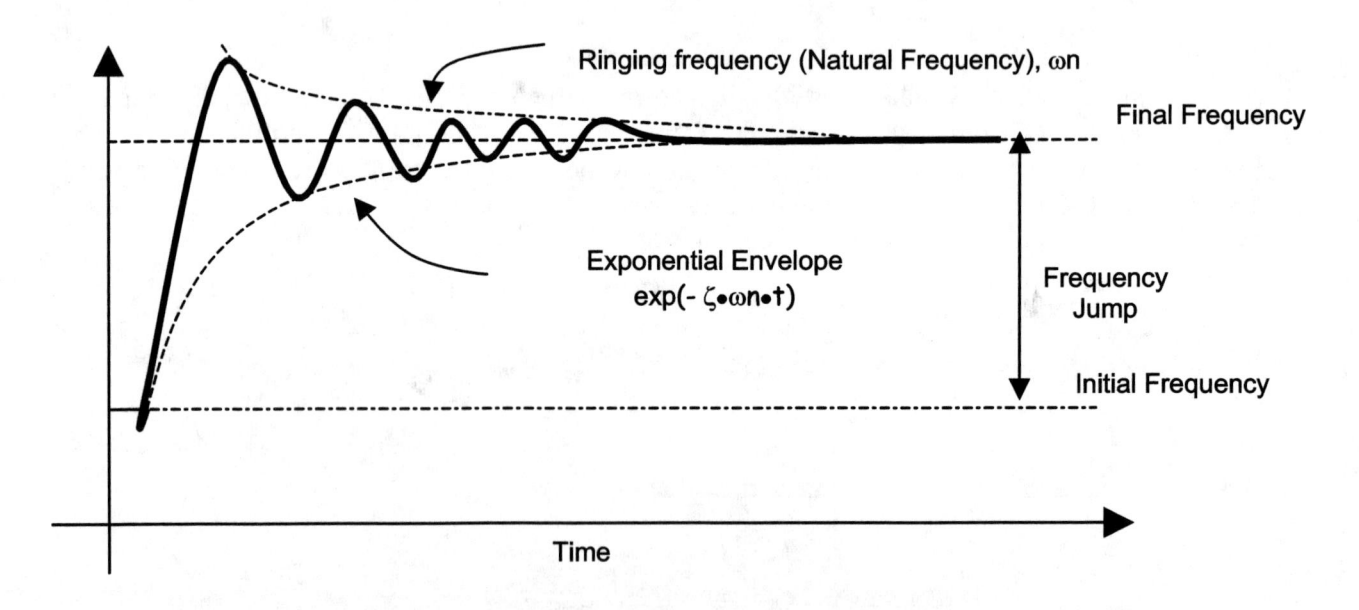

Figure 15.1 *Classical Model for the Transient Response of a PLL*

Relationship Between Phase Margin, Loop Bandwidth, Damping Factor, and Natural Frequency

For a second order filter, the following relationships exist for loop filters designed with National Semiconductor's AN-1001, National Semiconductor's online EasyPLL Program, or the equations presented in this book. These relationships are proven in the Appendix.

$$\omega c = 2 \cdot \zeta \cdot \omega n \tag{15.14}$$

$$\sec\phi - \tan\phi = \frac{1}{4 \cdot \zeta^2}$$

Phase Margin, ϕ	Damping Factor, ζ	Natural Frequency, ωn
30.00 degrees	0.6580	$0.7599 \cdot \omega c$
35.00 degrees	0.6930	$0.7215 \cdot \omega c$
36.87 degrees	0.7071	$0.7071 \cdot \omega c$
40.00 degrees	0.7322	$0.6829 \cdot \omega c$
45.00 degrees	0.7769	$0.6436 \cdot \omega c$
50.00 degrees	0.8288	$0.6033 \cdot \omega c$
55.00 degrees	0.8904	$0.5615 \cdot \omega c$
60.00 degrees	0.9659	$0.5177 \cdot \omega c$
61.93 degrees	1.0000	$0.5000 \cdot \omega c$
65.00 degrees	1.0619	$0.4709 \cdot \omega c$
70.00 degrees	1.1907	$0.4199 \cdot \omega c$

Table 15.1 *Relationship Between Phase Margin, Damping Factor and Natural Frequency*

So by specifying the loop bandwidth, ωc, and the phase margin, ϕ, the damping factor and natural frequency can be determined, and vise versa.

Fast Approximations to Lock Time, Rise Time, and Overshoot

If one assumes 50 degrees of phase margin, a settling tolerance of 1/100000 and factors in all the factors of $2 \cdot \pi$, the following simple relationships can be derived.

$$Lock\ Time \approx \frac{4}{Fc} \tag{15.15}$$

$$Peak\ Time \approx \frac{0.8}{Fc} \tag{15.16}$$

$$Overshoot \approx \frac{f2-f1}{9} \qquad (15.17)$$

These rules of thumb are easy to use and roughly accurate, except for the expression for overshoot. Because the approximation ignores the impact of *R2* and *C2*, the overshoot is much less. In practice it is more accurate to assume the overshoot is $1/3^{rd}$ of the frequency difference. Also note that the first two equations imply that the peak time is roughly one-fifth of the lock time.

Fourth Order Transient Analysis

This analysis considers all the poles and zeros of the transfer function and gives the most accurate results. To start with, the transfer function in (15.2) is multiplied by *(f2-f1)/(N•s)*. However, since these formulas are really referring to the phase response, and it is the frequency response that is sought, the whole transfer function is also multiplied by *s* to perform differentiation (frequency is the derivative of phase). The resulting expression is rewritten in the following form:

$$F(s) = s \bullet CL(s) \bullet \frac{f2-f1}{N \bullet s} = \frac{K \bullet (1+s \bullet T2)}{A3 \bullet s^5 + A2 \bullet s^4 + A1 \bullet s^3 + A0 \bullet s^2 + K \bullet T2 \bullet s + K} \qquad (15.18)$$

The challenge is finding the poles of the closed loop transfer function. The polynomial can be up to fifth order, depending on the loop filter order. Abel's Impossibility Theorem states that there cannot exist a closed form solution for polynomials of fifth and higher order. Closed form solutions do exist for polynomials of fourth and lower order, although the fourth and third order equations are rather complicated. If the means are not available to solve for the poles, then one can approximate by reducing the order of the polynomial, until it is solvable.

Note that the roots of the denominator correspond to the poles of the transfer function. Since this is a fourth order polynomial, the roots of this function can be found analytically, although it is much easier to find them numerically. The transient response can be rewritten as:

$$f(t) = \sum_{i=0}^{4^*} B_i \bullet \left[\frac{1}{s \bullet (s-p_i)} + \frac{T2}{s-p_i} \right] \qquad (15.19)$$

$$B_i = \frac{K \bullet (f2-f1)}{A3^*} \bullet \prod_{k \neq i} \frac{1}{p_i - p_k}$$

Note that if the filter order is lower than fourth order, or an approximation is being used, the terms with the asterisks will be reduced. For the above formulas and the formulas to follow, an asterisk will be used to designate places where the coefficient is changed depending on the filter order. For the above equations, for a fourth order filter, the summation index goes to 4 and the denominator for the coefficients is $A3$. For a third order filter, the summation index goes to 3 and the denominator for the coefficients is $A2$. For a second order filter, the summation index goes to 2 and the denominator for the coefficients is $A1$. Note that some of the coefficients B_i will be complex; however, they will combine in such a way that the final solution is real. Now since the poles need to be calculated for this, it will be assumed that they all have negative real parts. If this is not the case, then the design is unstable. Using this assumption that the design is stable, the transient response can be simplified. Also, if the simulator does not do this, the solution can be expressed with all real variables by applying Euler's formula:

$$e^{\alpha + j \bullet \beta} = e^{\alpha} \bullet (\cos \beta + j \bullet \sin \beta) \qquad (15.20)$$

Assuming a stable system, the transient response is:

$$f(t) = f2 + \sum_{i=0}^{4*} B_i \bullet e^{p_i \bullet t} \bullet \left(\frac{1}{p_i} + T2\right) \qquad (15.21)$$

Additional Comments Regarding the Lock Time Formula

Using the Exponential Envelope

(15.21) provides a complete analysis for the transient response, including all of the ringing of the PLL. However, for the purposes of lock time determination, it is better to eliminate the ringing from the equation, and study only the exponential envelope. This makes the prediction of lock time more consistent. The exponential envelope is obtained by applying the triangle inequality to (15.22).

$$\textbf{\textit{Exponential Envelope}} = f2 + \sum_{i=0}^{4*} \left| B_i \bullet e^{p_i \bullet t} \bullet \left(\frac{1}{p_i} + T2\right) \right| \qquad (15.22)$$

Cycle Slips

When an instantaneous phase error is presented to the phase detector, then cycle slipping can occur. When the N counter value changes, then the phase of the VCO signal divided by N will initially be incorrect in relation to the crystal reference signal divided by R. If the loop bandwidth is very small (around 1%) relative to the comparison frequency, then this phase error will accumulate faster than the PLL can correct for it and eventually cause the phase detector to put out a current correction of the wrong polarity. By dividing the comparison frequency by the instantaneous phase error presented to the phase detector, one can approximate how many cycles it would take the phase detector to cycle slip. If this time is less than about half the rise time of the PLL, then cycle slipping is likely to occur. An easier rule of thumb that is less accurate is that cycle slipping tends to occur when the loop bandwidth is less than 1% of the comparison frequency. Cycle slips are somewhat rare in integer PLL designs, but are common with fractional N PLL designs, since they typically run at higher comparison frequencies. Many of National Semiconductor's PLLs have features such as cycle slip reduction and Fastlock that reduce the effects of cycle slipping significantly or even completely.

Dependence of Lock Time on Loop Bandwidth

Consider a two loop filters that are designed for the exact same parameters, except for the second loop filter is designed to have a loop bandwidth of M times the loop bandwidth of the first filter. In this case, the scaling rule for loop filters apples. All the resistor values in the second filter will be M times the resistor values in the first filter and the capacitors values in the second filter will be $1/M^2$ times the capacitor values of the first filter. Substituting this in for the definition of the filter coefficients yields the result that $A0$ will be multiplied by $1/M^2$, $A1$ by $1/M^3$, $A2$ by $1/M^4$, and $A3$ by $1/M^5$. It therefore follows that if p which makes the denominator in equation (15.18) equal to zero for the first filter, then $M \bullet p$ will make the denominator in equation (15.18) equal to zero for the second filter. Combining this information with formula (15.19)yields the result that the coefficients B_i are divided by a factor of M. Looking at formulas (15.18) or (15.19), the factors of M all cancel out, except in the exponent. This proves that the transient response for the second loop filter will be identical to that of the first, except for the time axis is scaled by a factor of $1/M$. The grand result of all this analysis is that it proves that the lock time is inversely proportional to the loop bandwidth, and that the overshoot (undershoot) will remain exactly the same.

Dependence of Lock Time on the Frequency Jump

The quantity $|f2 - f1|$ is the frequency jump. Now consider the same loop filter. For the first lock time measurement, the transient response is recorded. For the second lock time measurement, the final frequency, $f2$, is kept constant, but the initial frequency, $f1$, is changed such that the frequency jump is increased by a factor of K, equation (15.18) and (15.19) will be the same for both cases, except for the fact that the coefficients for A_i in the second case will be multiplied by a factor of K. This implies that the transient response will be the same for both cases, except for in the second case, the ringing is multiplied by a factor of K. Note that although the lock time for the second case will be longer, it will be not be increased by a factor of K, but rather something much less. What can be implied from this is that if the frequency jump and frequency tolerance are scaled by equal amounts, the lock time will be identical.

Rule of Thumb for Lock Time for an Optimized Filter

Although (15.21) is very complete, it is difficult to apply without the aid of computers. Simulations show optimal lock time occurs with a phase margin around 48 degrees. Recall that it was shown that lock time was inversely proportional to loop bandwidth, and that the lock time does not change if the frequency jump and frequency tolerance are scaled in equal amounts. Using the above rules and assuming 48 degrees of phase margin, a rule of thumb for lock time can be derived from simulated data.

$$LT \approx \frac{400}{Fc} \cdot \left(1 - \log_{10}|\Delta F|\right) \quad (15.23)$$

$$\Delta F = \frac{Frequency\ Tolerance}{Frequency\ Jump}$$

LT is the lock time in microseconds, **Fc** is the loop bandwidth in kHz, and **ΔF** is the ratio as shown above.

Simulation Results

Figure 15.2 and Figure 15.3 show a comparison between simulated results, based on this chapter, and actual measured data. There is very good agreement between these graphs. Note that the *C2* capacitor in the loop filter was type C0G. When this was changed to a worse dielectric, the lock time increased from 489 µS to 578 µS. This example was also contrived so that the charge pump stayed away from the power supply rails, in order to eliminate the saturation effects of the charge pump. These are the effects that most often cause the measured result to differ from the theoretical result. The VCO capacitance was added to *C3* for the purposes of the calculations.

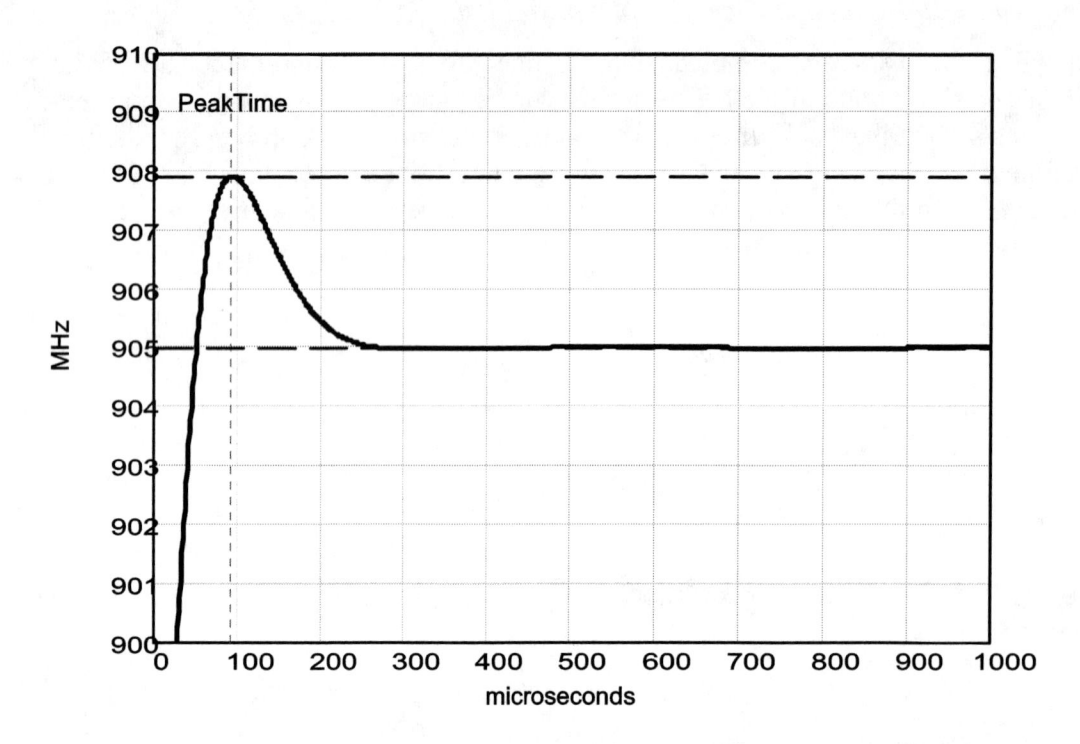

Figure 15.2 *Theoretical Peak Time of 94 μS to 907.9 MHz*

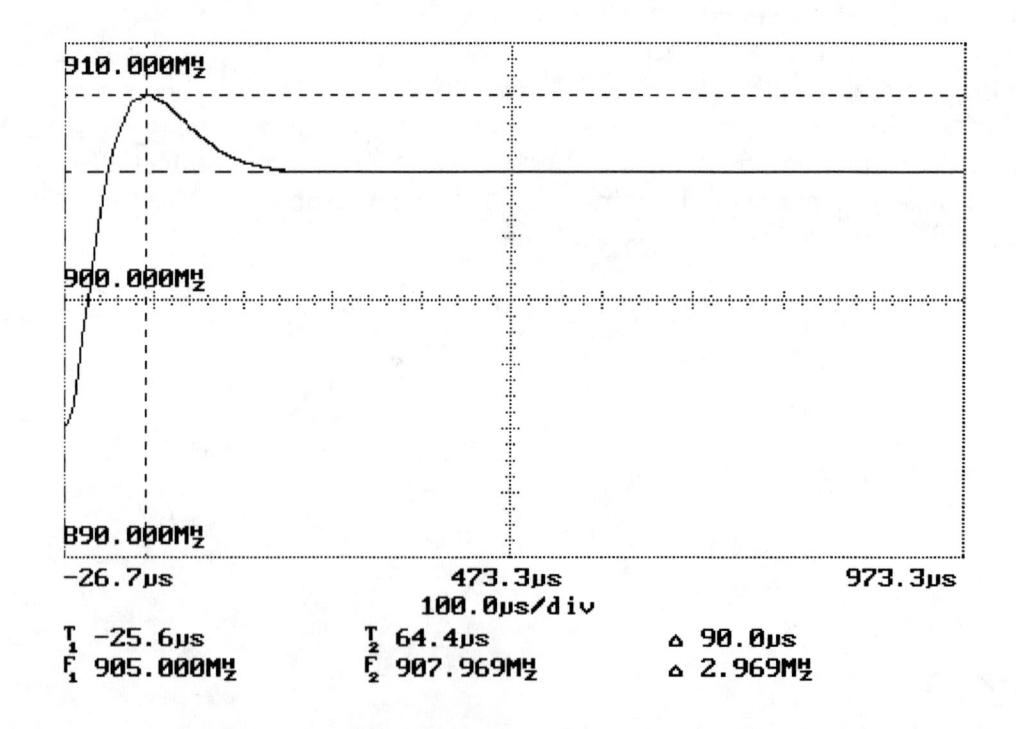

Figure 15.3 *Actual Peak Time of 90 μS to 908.0 MHz*

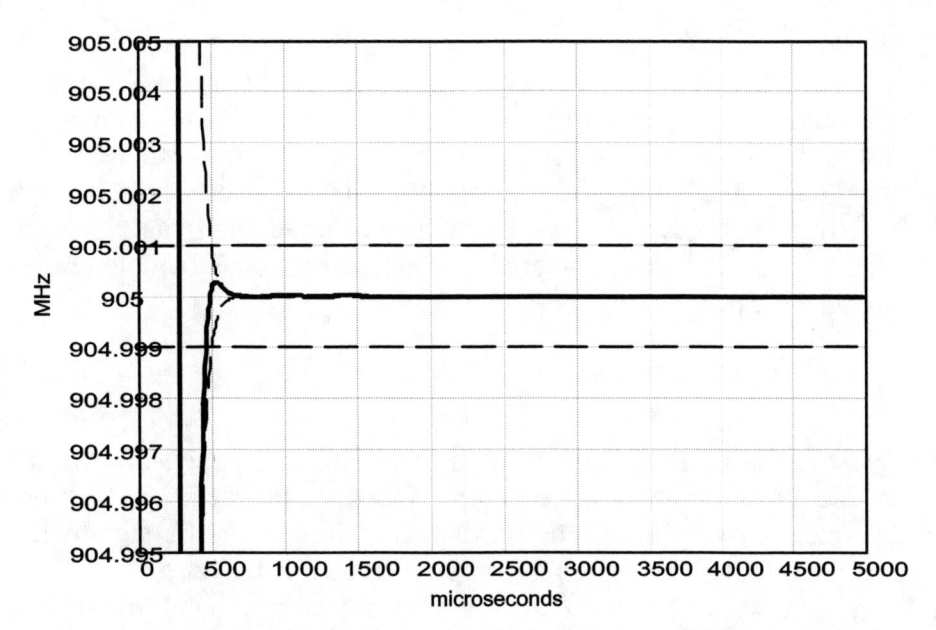

Figure 15.4 *Theoretical Lock Time to 1 kHz in 446 υS*

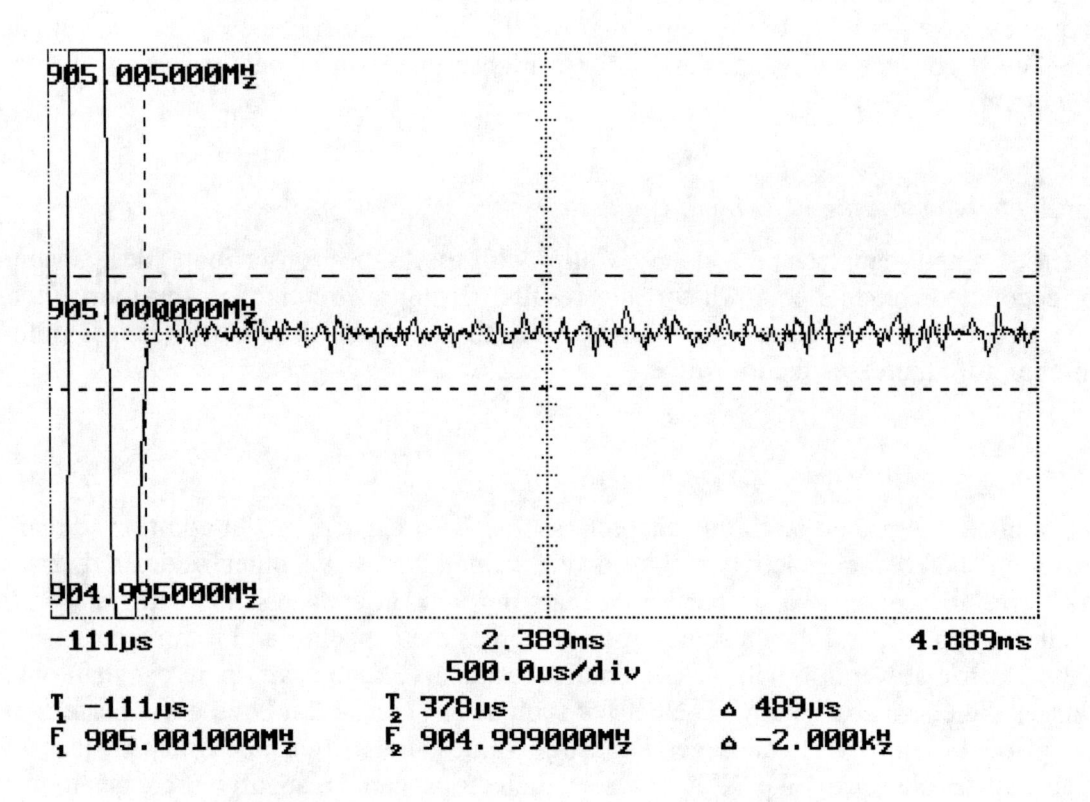

	T₁ -111µs	T₂ 378µs	Δ 489µs
	F₁ 905.001000MHz	F₂ 904.999000MHz	Δ -2.000kHz

Figure 15.5 *Actual Lock time to 1 kHz of 489 µS*

Conclusion

This chapter has gone through a rigorous derivation of the equations involved in predicting lock time and the transient response of the PLL when the *N* divider is changed. A second order and a fourth order model were presented. For the fourth order model, discrepancies between theoretical lock times and measured lock times are on the order of 10 - 20% or less. If theoretical lock times and measured lock times closely agree, then this indicates that this is the best the PLL can do. However, if there is a large discrepancy, then one or more of the factors below could be the cause.

VCO and Charge Pump Non-linearity

Perhaps the biggest real world effect that could throw off this analysis is the non-linear characteristics of the VCO and the charge pump. When switching from one frequency to another, there is typically overshoot in the order of one third of the frequency jump. This overshoot is dependent on the phase margin/damping factor. If the VCO overshoots too far past its intended range for usage, or if the tuning voltage ever gets too close (about 0.5 V) to the supply rails for the charge pump, the first lobe of the transient response gets longer and increases the lock time. The designer should be aware that if overshoot causes the frequency to go outside the tuning range of the VCO, the modeled prediction could lose accuracy. To deal with this, design for a higher phase margin in order to decrease the overshoot.

Not Accounting for the VCO Input Capacitance

The VCO input capacitance adds in parallel with the highest order loop filer capacitor. If not accounted for, this could distort the results. If not accounted for, the loop filter input capacitance tends to decrease the loop bandwidth and make the loop filter less optimized, which in turn increases the lock time.

Bad Capacitor Dielectrics

The simulations presented in this chapter assume ideal capacitors. In addition to real world capacitors not being exactly on the correct value, they have other undesired properties. Dielectric absorption is a property of capacitors. In order to test dielectric absorption, a voltage is applied, and then a short is placed across the capacitor and removed. Parts with a low dielectric absorption will have a smaller residual voltage develop across than ones with a larger dielectric absorption. Dielectrics such as NP0 and Film have good performance in this respect. However, for larger capacitor values, it is often necessary to use a lower performance dielectric like X7R. These dielectrics can drastically increase lock times. Some PLL designs seem completely immune to the impact of dielectrics, while others can have the lock time double or increase even more. If the actual lock time is substantially longer than the theoretical lock time, then replace the capacitors, especially capacitor *C2*, with ones of higher quality. For the example previously given, using a higher dielectric absorption capacitor for component *C2* increased the lock time from 489 µS to 578 µS.

Phase Detector Discrete Sampling Effects

The discrete sampling effects of the phase detector usually have little bearing on the lock time, provided that the comparison frequency is larger than about 10 times the loop bandwidth and less than 100 times the loop bandwidth. The next chapter on the discrete transient analysis discusses this in further detail.

Other Comments

There are also charge pump mismatch, charge pump leakage, board parasitics, and component leakages that could cause additional errors. Although the equations for a fourth order loop filter were not presented here, the transient response can be derived in a very similar way that was used to derive the transient response for the third order loop filter.

Appendix

The Relationship Between Natural Frequency (ωn), Damping Factor (ζ), Loop Bandwidth (ωc), and Phase Margin (ϕ)

Below is a list of equations that can be derived from the definitions for various parameters. The strategy for this derivation is to eliminate the parameters *T1*, *T2*, *A0*, and *A1* in order to find the desired relationship.

$$\gamma = \omega c^2 \bullet T2 \bullet \frac{A1}{A0} \approx 1 \tag{15.24}$$

$$\omega n = \sqrt{\frac{K\phi \bullet Kvco}{N \bullet A0}} \tag{15.25}$$

$$\zeta = \frac{T2}{2} \bullet \omega n \tag{15.26}$$

$$\frac{K\phi \bullet Kvco}{N \bullet \omega c^2 \bullet A0} \bullet \frac{\sqrt{1 + \omega c^2 \bullet T2^2}}{\sqrt{1 + \omega c^2 \bullet T1^2}} = 1 \tag{15.27}$$

$$\pi - arctan(\omega c \bullet T2) + arctan(\omega c \bullet T1) = \phi \tag{15.28}$$

$$\frac{A1}{A0} = T1 \tag{15.29}$$

Eliminating *A1* and *A0* yields the following new equations:

$$1 = \omega c^2 \bullet T2 \bullet T1 \tag{15.30}$$

$$\zeta = \frac{T2}{2} \bullet \omega n \tag{15.31}$$

$$\frac{\omega n^2}{\omega c^2} \bullet \frac{\sqrt{1 + \omega c^2 \bullet T2^2}}{\sqrt{1 + \omega c^2 \bullet T1^2}} = 1 \tag{15.32}$$

$$T1 = \frac{sec\phi - tan\phi}{\omega c} \tag{15.33}$$

$T2$ and $T1$ can be eliminated as follows:

$$T2 = \frac{2 \cdot \zeta}{\omega n} \tag{15.34}$$

$$T1 = \frac{\omega n}{2 \cdot \zeta \cdot \omega c^2} \tag{15.35}$$

These values can be substituted in:

$$\left[4 \cdot \zeta^2 \cdot \left(\frac{\omega n}{\omega c}\right)^2 - 1\right] + \left[\left(\frac{\omega n}{\omega c}\right)^4 - \frac{1}{4 \cdot \zeta^2} \cdot \left(\frac{\omega n}{\omega c}\right)^2\right] = 0 \tag{15.36}$$

By inspection, both terms will be zero provided that:

$$\frac{\omega n}{\omega c} = \frac{1}{2 \cdot \zeta} \quad \Rightarrow \quad \omega c = 2 \cdot \zeta \cdot \omega n \tag{15.37}$$

Substituting in this new result yields the other relationship:

$$\sec\phi - \tan\phi = \frac{1}{4 \cdot \zeta^2} \tag{15.38}$$

Chapter 16   Discrete Lock Time Analysis

Introduction

The previous model for lock time assumes that the charge pump puts out a continuous current that is proportional to the phase error. In reality, the charge pump actually outputs a pulse width modulated signal. Modeling this as an analog signal usually serves as a good approximation provided that the loop bandwidth of the PLL is between one-tenth and one-hundredth of the comparison frequency. The loop bandwidth is usually kept less than one-tenth of the comparison frequency as a design requirement. However, there is no design requirement that drives the loop bandwidth to be more than one-hundredth of the comparison frequency. This condition is often violated in the case of fractional PLLs. This chapter models the lock time in a discrete fashion and investigates some of the discrete effects of the charge pump on lock time.

Modeling the Lock Time

The discrete lock time model for lock time is well suited for computer modeling because it creates a set of difference equations. In order to derive this model, the following steps are taken:

- Define all voltages across the capacitors
- Derive the differential equations involved
- Convert the differential equations to difference equations
- Solve the system by incrementing in small discrete time steps

Deriving the Nomenclature

Although a trivial step, defining the problem in the right way simplifies the analysis. The easiest convention to use is define all the voltages to be across the capacitors, and all the currents to be through the capacitors. For instance, V*C1* stands for the voltage across capacitor *C1*, and *i1* stands for the current through capacitor *C1*. Once this is done, the equations are easy to derive.

Deriving the Equations

To derive the equations, the first step is to initialize all voltages to zero. Define the tuning voltage such that zero volts corresponds to the initial frequency. Then derive equations to calculate the new change in voltages from the old voltages. Add these to the old voltages in order to compute the new voltages. Then calculate the new VCO frequency, and then finally the new phase. The new phase can be calculated by adding the product of the time step times the frequency. It is best to think of this phase in cycles. The charge pump state can then be known, and then the process repeats itself.

Step 1:   *Initialize all States*

Define all states and voltages to be zero. Define a time increment, which should be much smaller than the period of the comparison frequency. Define the frequency of the VCO to be the starting frequency.

Step 2:   *Determine the new charge pump state and current*

Define this as *ICPout*. A charge pump event occurs when the phases of one of the inputs to the phase detector exceeds one and it causes the phase detector to change states

Step 3:   *Determine the new voltage at the VCO*

For example, in a second order filter the following equations hold:

$$I_{CPout} = i1 + i2 \qquad (16.1)$$
$$i1 = \frac{1}{C1} \cdot \frac{\Delta VC1}{\Delta t}$$
$$i2 = \frac{1}{C2} \cdot \frac{\Delta VC2}{\Delta t}$$
$$VC2 + i2 \cdot R2 = VC1$$

Now these equations can be combined to solve for $\Delta VC1$ and $\Delta VC2$. Once these are known, the new VCO frequency can be found. The table below shows the values for various orders of active and passive loop filters.

		Filter Type	
		Passive	Active
Filter Order	2nd	$\Delta VC2 = \dfrac{\Delta t \cdot (VC1-VC2)}{C2 \cdot R2}$ $\Delta VC1 = \dfrac{\Delta t \cdot CPout}{C1} - \dfrac{C2 \cdot \Delta VC2}{C1}$	
	3rd	$\Delta VC3 = \dfrac{\Delta t \cdot (VC1-VC3)}{C3 \cdot R3}$ $\Delta VC2 = \dfrac{\Delta t \cdot (VC1-VC2)}{C2 \cdot R2}$ $\Delta VC1 = \dfrac{\Delta t \cdot CPout}{C1} - \dfrac{C2 \cdot \Delta VC2}{C1}$	$\Delta VC3 = \dfrac{\Delta t \cdot AMP \cdot (VC1-VC3)}{C3 \cdot R3}$ $\Delta VC2 = \dfrac{\Delta t \cdot (VC1-VC2)}{C2 \cdot R2}$ $\Delta VC1 = \dfrac{\Delta t \cdot CPout}{C1} - \dfrac{C2 \cdot \Delta VC2}{C1}$
	4th	$\Delta VC4 = \dfrac{\Delta t \cdot (VC3-VC4)}{C4 \cdot R4}$ $\Delta VC3 = \dfrac{\Delta t \cdot (VC1-VC3)}{C3 \cdot R3} - \dfrac{C4 \cdot \Delta VC4}{C3}$ $\Delta VC2 = \dfrac{\Delta t \cdot (VC1-VC2)}{C2 \cdot R2}$ $\Delta VC1 = \dfrac{\Delta t \cdot CPout}{C1} - \dfrac{C4 \cdot \Delta VC4}{C1} - \dfrac{C3 \cdot \Delta VC3}{C1} - \dfrac{C2 \cdot \Delta VC2}{C1}$	$\Delta VC4 = \dfrac{\Delta t \cdot (VC3-VC4)}{C3 \cdot R3}$ $\Delta VC3 = \dfrac{\Delta t \cdot (AMP \cdot VC1-VC3)}{C3 \cdot R3} - \dfrac{C4 \cdot \Delta VC4}{C4 \cdot R4}$ $\Delta VC2 = \dfrac{\Delta t \cdot (VC1-VC2)}{C2 \cdot R2}$ $\Delta VC1 = \dfrac{\Delta t \cdot CPout}{C1} - \dfrac{C2 \cdot \Delta VC2}{C1}$

Figure 16.1 *Discrete Lock Time Formulae*

Step 4: Calculate the new VCO frequency

The tuning voltage will be the voltage across the highest order capacitor.

Step 5: Calculate the New Phases for the Inputs of the Phase Detector

Recalling that 1 Hz is one cycle per second, calculated the fraction of added cycles by multiplying the frequency times the time step and adding it to the current number of cycles. If this cycle exceeds one, consider a charge pump event. Now return to step 2.

Comments Regarding Computational Accuracy

The accuracy of the computations are limited by the size of the time step Δt. This typically requires a time step that is too small to be practical. However, the transient response that happens up to the peak time is the part that is most impacted by the discrete sampling effects of the charge pump. This portion of the transient response is of the most interest when studying discrete sampling effects and is also much less sensitive to the step size. Using solution methods like Runga-Kutta do not really improve the accuracy because the limiting factor is that there should be at least 8 time cycles within the amount of time the charge pump comes on in order to get a good final settling frequency tolerance. One trick to improve this is to make the size of the time step dynamic such that the resolution is finer near the times the charge pump is on. Another good trick is to bail out of the routine once the frequency is close and there are less than 8 time steps in one charge pump event. If one studies the analog simulation, the increase in lock time due to discrete sampling effects will be roughly equal to the increase in the peak time due to discrete sampling effects.

Cycle Slipping

Cause of Cycle Slipping

The cause of cycle slipping is that the charge pump goes from a very high duty cycle to a very low duty cycle. The charge pump does not pump in the wrong direction in order to cause a cycle slip. What happens is that a large voltage is developed across the resistor *R2* in the loop filter when the charge pump current is flowing, and when it is removed, there is a corresponding drop in the VCO tuning voltage. Note that the capacitor *C1* and the other loop filter components may reduce this voltage drop. In the example below, a single cycle slip occurs around 17 µS. In this particular case, the cycle slip has only a small impact on the lock time. However, in the above example, there can be far more cycle slips that can greatly degrade the lock time.

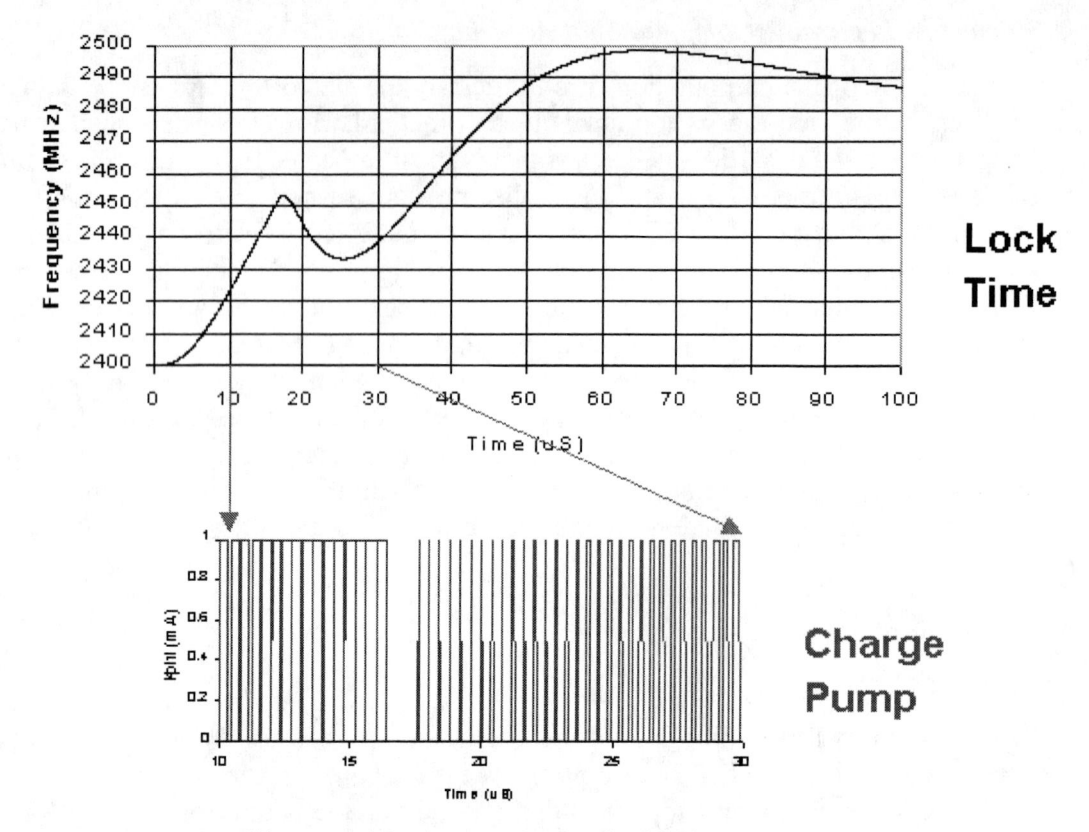

Figure 16.2 *Anatomy of a Cycle Slip*

Calculating if Cycle Slipping is an Issue

Assuming that both the *N* and the *R* counters start off in phase and the loop bandwidth is infinite, the time to the first cycle slip is when one the *N* counter gets an extra cycle.

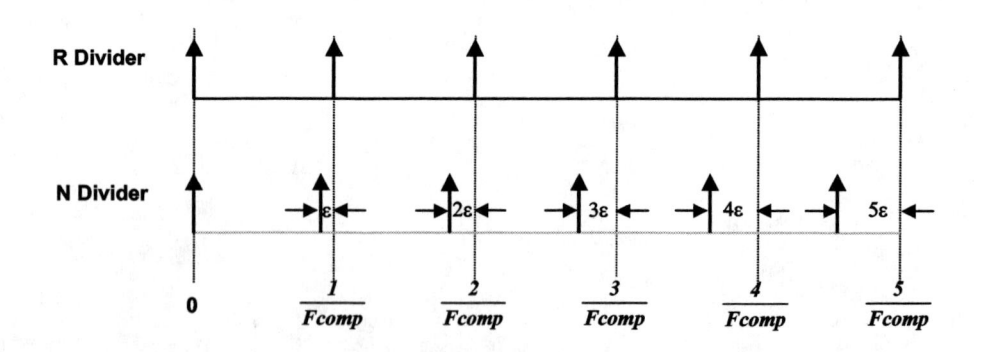

Figure 16.3 *Calculating the time to the First Cycle Slip for an Infinite Loop Bandwidth*

This time to the first cycle slip with a zero Hz loop bandwidth is:

$$t = \frac{1}{Fcomp} \cdot \frac{\left(\frac{1}{Fcomp}\right)}{\varepsilon} = \frac{1}{Fcomp^2 \cdot \left|\frac{N}{f1} - \frac{N}{f2}\right|} = \frac{1}{Fcomp \cdot \left|1 - \frac{f1}{f2}\right|} \qquad (16.2)$$

Now if the PLL is in lock, no cycle slipping occurs and t is infinite. However, if not, this time should be about the rise time or more in order to avoid cycle slipping. In actuality, the loop bandwidth is not infinite, and assuming that no cycle slipping occurs before the peak time is way to conservative. A more reasonable assumption is to assume that the first cycle slip cannot occur before one-fourth of the peak time. Applying this rule and using the equations for peak time from the previous chapter give the following relationship.

$$\frac{Fcomp}{BW} < \frac{5}{\left|1 - \frac{f2}{f1}\right|} \qquad (16.3)$$

For instance, if the frequency changes 5%, then the ratio of the comparison frequency to the loop bandwidth should be no more than 100 if one wants to avoid the effects of cycle slipping. This factor of 100 will be used throughout this book for the sake of simplicity.

Impact of Cycle Slipping on Lock Time

But what is the impact on lock time if this limit is exceeded? In general, this can be approximated by the increase in the peak time times the overdrive factor. The overdrive factor can never be less than one and the factor by which this limit is exceeded.

$$Overdrive\ Factor = \max\left\{\frac{2}{3} \times \frac{\left(Fcomp/BW\right)}{\left(\frac{5}{\left|1 - \frac{f2}{f1}\right|}\right)}, 1\right\} \qquad (16.4)$$

$$Peak\ Time(Accounting\ for\ Discrete\ Effects) = Overdrive\ Factor \times \frac{0.8}{BW} \qquad (16.5)$$

Now the factor of 2/3 comes in the overdrive factor comes in for two reasons. The first is that the overshoot typically is a lot less in cases where cycle slipping is occurring and this decreases the peak time. The second reason is that the severity of the cycle slipping decreases as the PLL gets closer to the target frequency. There is nothing magical about the number 2/3 and this number comes from experience and simulation, as opposed to mathematical derivation.

Figure 16.4 shows the lock time for a PLL system with a 2 kHz loop bandwidth and various comparison frequencies. Note that when the comparison frequency is less than 100 times the loop bandwidth (200 kHz in this case), the analog and discrete lock time model agree. However, when the comparison frequency is much larger than 100 times the loop bandwidth, the rise time is greatly increased, which in turn increases the lock time. It is a reasonably accurate rule of thumb to assume that the amount that the lock time is increased due to cycle slipping is equal to the amount that the rise time is increased by cycle slipping.

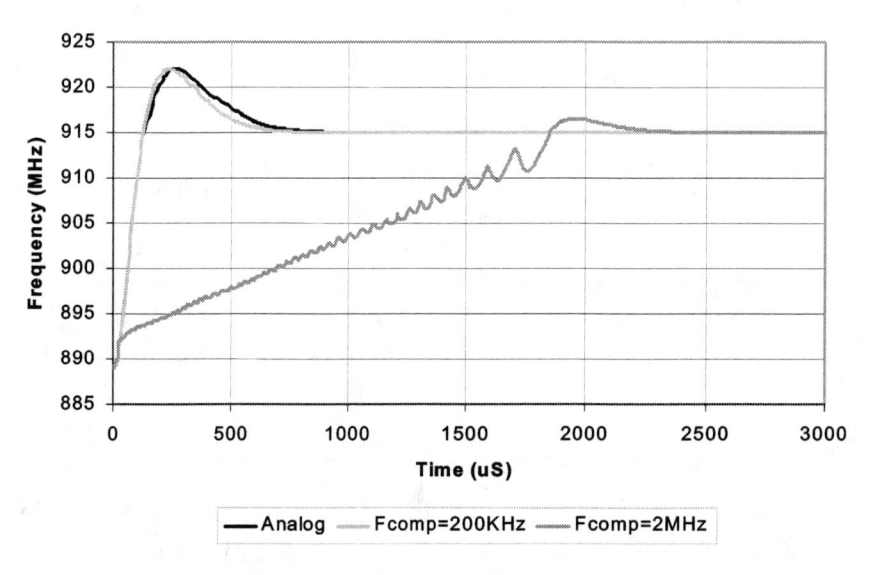

Figure 16.4 *Cycle Slip Example*

Conclusion

The analog model of lock time does a good job provided that the comparison frequency does not exceed about 100 times the loop bandwidth and is not less than 10 times the loop bandwidth. There are many advantages of the analog method including computational speed and accuracy. However, in situations where the comparison frequency exceeds 100 times the loop bandwidth, discrete sampling effects become relevant and the most significant impact is the increase in lock time. In situations where the comparison frequency is too small compared to the loop bandwidth, the impact of the discrete sampling action of the phase detector sometimes increases lock time and sometimes degrades it. If the comparison frequency is not at least five times the loop bandwidth, instability is very likely to happen.

Chapter 17   Routh Stability for PLL Loop Filters

Introduction

There are two ways to make a loop filter unstable. The first is to design for a loop bandwidth that is more than about $1/3^{rd}$ of the comparison frequency. The second is to design a loop filter such that the poles of the closed loop system fall in the right hand plane. This can happen when the phase margin is too low, at least for a third order filter. For the purposes of this chapter, the term Routh stability refers to a system where all the poles of the closed loop transfer function are in the left hand plane. This chapter examines what restrictions Routh's Stability Criterion implies.

Calculation of Stability Coefficients

The open loop transfer function for a loop filter up to 4th order can be expressed as follows:

$$G(s) = \frac{N \bullet K \bullet (1 + s \bullet T2)}{s^2 \bullet (A3 \bullet s^3 + A2 \bullet s^2 + A1 \bullet s + A0)} \qquad (17.1)$$

$$K = \frac{K\phi \bullet Kvco}{N} \qquad (17.2)$$

The closed loop transfer function is as follows:

$$\frac{G(s)}{1 + G(s)/N} = \frac{N \bullet K \bullet (1 + s \bullet T2)}{A3 \bullet s^5 + A2 \bullet s^4 + A1 \bullet s^3 + A0 \bullet s^2 + T2 \bullet K \bullet s + K} \qquad (17.3)$$

Formation of a Routh Table

The system will be stable if all of the poles of the denominator have negative real parts. Instead of explicitly calculating the roots, it is far easier to use Routh's stability criterion, which says that all the roots have negative real parts if and only if the elements in the Routh array are positive. The elements in the Routh Array are the elements in the second column of the Routh table that is shown below. The Routh table is formed by putting the odd terms in the first row and the even terms in the second row. Note that the term with the highest power is considered to be the first term, and therefore an odd term. The lower rows are formed by taking the determinant of the 2 X 2 matrix formed by eliminating the column that the entry of interest is in, and dividing by the first entry in the row above the entry of interest. Any row can be multiplied by a positive constant without affecting stability. Since all the filter coefficients are positive, this means that the denominator portions of the formulas may be disregarded.

s^n	d_n	d_{n-2}	d_{n-4}	...
s^{n-1}	d_{n-1}	d_{n-3}	d_{n-5}	...
	$b_1 = \dfrac{d_{n-1} \cdot d_{n-2} - d_n \cdot d_{n-3}}{d_{n-1}}$	$b_2 = \dfrac{d_{n-1} \cdot d_{n-4} - d_n \cdot d_{n-5}}{d_{n-1}}$	...	...
	$c_1 = \dfrac{b_1 \cdot d_{n-3} - b_2 \cdot d_{n-1}}{b_1}$	...	...	...

Table 17.1 *A Generic Routh Table*

Proof of Routh Stability for a Second Order Filter

The second order loop filter is a special case where $A3 = A2 = 0$.

s^3	$A1 = T1 \cdot A0$	$T2 \cdot K$
s^2	$A0$	K
	$K \cdot A0 \cdot (T2 - T1)$	0
	K	0

Table 17.2 *Routh Table for Second Order Loop Filter*

Now from the definition of K, it is clear that $K>0$. From the third row, this puts the restriction that $T2 > T1$. For a second order filter, this is always the case because:

$$T2 = R2 \cdot C2 \tag{17.4}$$

$$T1 = T2 \cdot \frac{C1}{C1 + C2}$$

Although not shown, in the case that the alternative feedback approach is used with an OP AMP, $T2>T1$ is also always true.

Theorem 1:

Using real non-zero component values and the standard loop filter topology, it is impossible to design a second order loop filter which is unstable, provided that the loop bandwidth is sufficiently small to justify the continuous time approximation.

So using the standard topology, it is impossible to design a loop filter that is unstable due to too low phase margin or poles in the right hand plane. This stability makes the second order filter a good choice when the VCO gain, charge pump gain, or N value drastically varies.

Conditions for Third Order Routh Stability

For the third order filter, it turns out that the Routh table is not so simple and that it is possible to design an unstable loop filter, regardless of loop bandwidth. Since the loop bandwidth decreases as the charge pump gain or VCO gain decreases, reducing these will eventually guarantee second order filter stability, and will always make a third order filter stable provided $T2 > T1 + T3$. For the purposes of simplifying the math in the Routh table, the following constant is introduced.

s^4	$A2$	$A0$	K
s^3	$A1$	$K \cdot T2$	0
	$A1 \cdot A0 - K \cdot A2 \cdot T2 = x$	$K \cdot A1$	0
	$K \cdot (T2 \cdot x - A1^2)$	0	0
	$K^2 \cdot A1 \cdot (T2 \cdot x - A1^2)$	0	0

Table 17.3 *Third Order Routh Stability Table*

The closed loop system will be stable provided that

$$T2 \cdot x - A1^2 > 0 \qquad (17.5)$$

If this is expressed in terms of filter coefficients, then the following rule can be derived:

$$T2 - T1 - T3 > \frac{K \cdot T2^2 \cdot T1 \cdot T3}{A0 \cdot (T1 + T3)} \qquad (17.6)$$

This criteria implies that $T2 > T1 + T3$ and there is some maximum upper bound for K. Since K includes the charge pump gain, VCO gain, and N divider, and $T2 > T1 + T3$ is a constraint applied in all loop passive loop filter designs, this implies that a passive third order loop filter can be made stable by sufficiently increasing N, sufficiently decreasing the VCO gain, or sufficiently decreasing the charge pump gain.

Conditions for Fourth Order Routh Stability

For the fourth order filter, there is some added complexity, but the general rule remains the same. There is a restriction on high the loop gain, **K**, can be, and also there is a restriction that *T2 > T1 + T3 + T4*. Table 17.4 shows the coefficients for a fourth order loop filter.

s^5	A3	A1	K•T2
s^4	A2	A0	K
s^3	A2•A1 - A0•A3 = x	K•(A2•T2 – A3)= y	0
	A0•x - A2•y	K•x	0
	y•(A0•x - A2•y) - K•x^2	0	0
	K•[y•(A0•x - A2•y) - K•x^2]	0	0

Table 17.4 *Fourth Order Routh Stability Table*

This imposes three constraints:

$$A2 \cdot A1 - A0 \cdot A3 > 0 \tag{17.7}$$

$$A0 \cdot x - A2 \cdot y > 0 \tag{17.8}$$

$$y \cdot z - K \cdot x^2 > 0 \tag{17.9}$$

If one substitutes in the time constants in place of the filter coefficients, we find that the first constraint is always satisfied. The second constraint implies:

$$\frac{T2}{\frac{1}{T1} + \frac{1}{T3} + \frac{1}{T4}} > 1 \tag{17.10}$$

The third constraint implies that:

$$y \cdot A0 - A2 \cdot y^2 - K \cdot x^2 > 0 \tag{17.11}$$

$$\Rightarrow \frac{K \cdot (A2 \cdot T2 - A3)}{x} < A2 \cdot (T2 \cdot A0 - A1)$$

Substituting in the values for the poles and recalling that *x>0* reduces this constraint to the following:

$$T2 > T1 + T3 + T4 \tag{17.12}$$

As in the case of a third order filter, this also implies that there is some restriction on the loop gain constant as well. Note that although it is possible for a fourth order loop filter to have complex poles, the sum of these poles will always be real.

Conclusion

The conditions for stability of loop filters have been investigated. There is always the condition that the loop bandwidth be sufficiently narrow relative to the comparison frequency, but there is also the constraint that all the poles of the closed loop transfer function have negative real parts. For the second order filter, this was shown to always be the case, but for the third and fourth order loop filters, there were real restrictions. In addition to always being Routh stable, the second order loop filter also tends to be the most resistant to variations the N counter value, charge pump gain, and VCO gain.

This chapter was actually inspired by the quest to find a filter that attenuated the spurs more. Notice that *T2* must be larger than *T1* or *T3* for the PLL to be stable. Theoretically, if *T3* or *T1* is chosen larger than *T2*, then the spurs could be reduced significantly. This chapter on Routh Stability proves why this type of loop filter will never be stable. The zero *T2* is necessary for stability because of the *1/s* factor introduced by the VCO.

References

Franklin, G., et. al.   *Feedback Control of Dynamic Systems*   Addison Wesley

Chapter 18   A Sample PLL Analysis

Introduction

This chapter is an example of a PLL analysis that applies many of the formulas and concepts that were derived in previous chapters.

Symbol	Description	Value	Units
$K\phi$	Charge Pump Gain	5	mA
$Kvco$	VCO Gain	30	MHz/V
$Fout$	Output Frequency	900	MHz
$Fcomp$	Comparison Frequency	200	kHz
$C1$	Loop Filter Capacitor	5.600	nF
$C2$	Loop Filter Capacitor	100.00	nF
$C3$	Loop Filter Capacitor	0.330	nF
$C4*$	Loop Filter Capacitor *(Not Accounting For VCO input Capacitance)	0.082	nF
$VCOcap$	VCO Input Capacitance	0.022	nF
$R2$	Loop Filter Resistor	1.0	kΩ
$R3$	Loop Filter Resistor	6.8	kΩ
$R4$	Loop Filter Resistor	33.0	kΩ

Calculate Basic Parameters

$$N = \frac{Fout}{Fcomp} \tag{18.1}$$

$$C4 = C4* + VCOcap \tag{18.2}$$

$$A0 = C1 + C2 + C3 + C4 \tag{18.3}$$

$$A1 = C2 \bullet R2 \bullet (C1+C3+C4) + R3 \bullet (C1+C2) \bullet (C3+C4) + C4 \bullet R4 \bullet (C1+C2+C3) \tag{18.4}$$

$$A2 = C1 \bullet C2 \bullet R2 \bullet R3 \bullet (C3+C4) + C4 \bullet R4 \bullet (C2 \bullet C3 \bullet R3 + C1 \bullet C3 \bullet R3 + C1 \bullet C2 \bullet R2) \tag{18.5}$$

$$A3 = C1 \bullet C2 \bullet C3 \bullet C4 \bullet R2 \bullet R3 \bullet R4 \tag{18.6}$$

Define

$$Z(s) = \frac{1 + s \cdot C2 \cdot R2}{s \cdot (A3 \cdot s^3 + A2 \cdot s^2 + A1 \cdot s + A0)} \quad (18.7)$$

$$G(s) = \frac{K\phi \cdot Kvco \cdot Z(s)}{s}$$

Fc, the loop bandwidth can be found by numerically solving the following equation.

$$|G(Fc \cdot 2\pi \cdot i)| = N \quad (18.8)$$

Once Fc is known, the phase margin and gamma optimization parameter can also be calculated.

$$\phi = \angle G(Fc \cdot 2\pi \cdot i) + 180° \quad (18.9)$$

$$\gamma = \frac{(Fc \cdot 2\pi)^2 \cdot C2 \cdot R2 \cdot A1}{A0} \quad (18.10)$$

Here are the results so far.

Symbol	Description	Value	Units
N	N Counter Value	4500	none
$C4$	Loop Filter Capacitor accounting for VCO input Capacitance	0.104	nF
$A0$	Total Capacitance	106.0340	nF
$A1$	First order loop filter coefficient	1.2786×10^{-3}	nFs
$A2$	Second Order loop filter coefficient	4.3879×10^{-9}	nFs^2
$A3$	Third Order loop filter coefficient	4.3128×10^{-15}	nFs^3
Fc	Loop Bandwidth	5.0813	kHz
ϕ	Phase Margin	50.7796	degrees
γ	Gamma Optimization Parameter	1.2292	none

Now Find the Poles and Zero

Solving for $T1$ is the hard part. First a cubic polynomial must be solved, and the results need to be manipulated in order to obtain $T1$. Once $T1$ is known, the other poles are easy to find. The following cubic polynomial needs to be solved for x, and then y can be calculated.

$$x^3 - 2 \cdot \frac{A1}{A0} \cdot x^2 + \left(\frac{A1^2}{A0^2} + \frac{A2}{A0}\right) \cdot x + \left(\frac{A3}{A0} - \frac{A1 \cdot A2}{A0^2}\right) = 0 \quad (18.11)$$

$$y = x^2 - \frac{A1}{A0} \cdot x + \frac{A2}{A0} \qquad (18.12)$$

$$T3, T4 = \frac{x \pm \sqrt{x^2 - 4 \cdot y}}{2} \qquad (18.13)$$

$$T1 = \frac{A3}{A0 \cdot y} \qquad (18.14)$$

Now that the poles are known, reorder them such that:

$$T1 \geq T3 \geq T4 \qquad (18.15)$$

Calculate the Zero

$$T2 = C2 \cdot R2 \qquad (18.16)$$

Symbol	Description	Value	Units
x	Intermediate Calculation	1.0329×10^{-5}	s
y	Intermediate Calculation	2.3518×10^{-11}	s^2
$T1$	Loop Filter Pole	6.9403×10^{-6}	s
$T2$	Loop Filter Zero	1.0000×10^{-4}	s
$T3$	Loop Filter Pole	3.3887×10^{-6}	s
$T4$	Loop Filter Pole	1.7294×10^{-6}	s
$\dfrac{T3}{T1}$	Pole Ratio	48.8264	%
$\dfrac{T4}{T3}$	Pole Ratio	51.0360	%
$\dfrac{1}{2\pi \cdot \|T1\|}$	Frequency of Loop Filter Pole	22.9321	kHz
$\dfrac{1}{2\pi \cdot T2}$	Frequency of Loop Filter Zero	1.5915	kHz
$\dfrac{1}{2\pi \cdot \|T3\|}$	Frequency of Loop Filter Pole	46.9667	kHz
$\dfrac{1}{2\pi \cdot \|T4\|}$	Frequency of Loop Filter Pole	92.0265	kHz

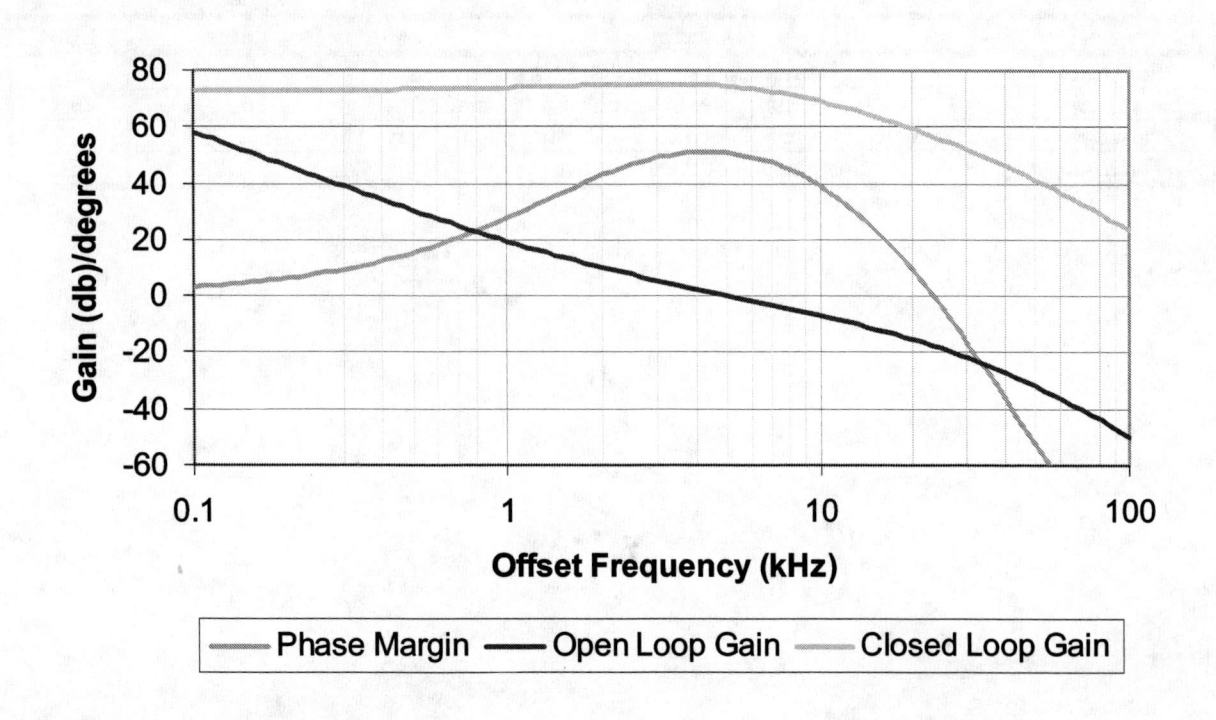

Reference Spur Simulation

Symbol	Description	Value	Units
BaseLeakageSpur	Constant for All Leakage Based Spurs	16.0	dBc
BasePulseSpur	Base Pulse Spur for the LMX2331U	-311	dBc
LeakageCurrent	Charge Pump Leakage Current	1	nA

Calculate the spur levels. Recall that *G(s)* has already been defined earlier.

$$SpurGain = 20 \cdot \log |G(Fcomp \cdot 2\pi \cdot i)| \qquad (18.17)$$

$$LeakageSpur = BaseLeakageSpur + 20 \cdot \log \left| \frac{LeakageCurrent}{K\phi} \right| + SpurGain \qquad (18.18)$$

$$PulseSpur = BasePulseSpur + SpurGain + 40 \cdot \log \left| \frac{Fcomp}{1\,Hz} \right| \qquad (18.19)$$

$$TotalSpur = 10 \cdot \log \left| 10^{\frac{LeakageSpur}{10}} + 10^{\frac{PulseSpur}{10}} \right| \qquad (18.20)$$

Symbol	Description	Value	Units
SpurGain	Spur Gain for this spur at 200 kHz offset	1.7663	dB
LeakageSpur	Estimated spur level based only on leakage	-116.2131	dBc
PulseSpur	Estimated spur level based only on charge pump pulse	-97.1925	dBc
TotalSpur	Combination of the above two spurs	-97.1385	dBc

Phase Noise Simulation

Symbol	Description	Value	Units
$K\phi Knee$	Phase noise knee current	1000	µA
$PN1Hz^*$	1 Hz Normalized Phase Noise for infinite $K\phi$	-214.8	dBc/Hz
$PN1Hz$	Calculated 1 Hz Normalized Phase Noise for this $K\phi$	-214.0	dBc/Hz
$PN10kHz$	1 GHz Normalized 1/f Noise @ 10 kHz offset	-94.6	dBc/Hz
$Plateau$	Plateau frequency for 1/f noise	1.00	kHz
$VCO10kHz$	VCO phase noise @ 10 kHz offset	-105	kHz
$TCXO10khz$	TCXO frequency @ 10 kHz offset	-160	dBc/Hz
$TCXO$	Crystal Frequency	10	MHz

PLL Noise

Flat Noise

$$PN1Hz = PN1Hz^* + 10 \cdot \log\left|1 + \frac{K\phi Knee}{K\phi}\right| \qquad (18.21)$$

$$PN\_Flat(f) = PN1Hz + 10 \cdot \log\left|\frac{Fcomp}{1Hz}\right| \qquad (18.22)$$

1/f Noise

$$PN\_Slope(f) = PN10kHz + 20 \cdot \log\left|\frac{Fout}{1GHz}\right| - 10 \cdot \log\left|\frac{\max\{f, Plateau\}}{10kHz}\right| - 20 \cdot \log|N| \qquad (18.23)$$

Total PLL Noise

$$Noise\_PLL(f) = 10 \cdot \log\left|10^{PN\_Flat(f)/10} + 10^{PN\_Slope(f)/10}\right| + 20 \cdot \log\left|\frac{G(f \cdot 2\pi \cdot i)}{1 + \frac{G(f \cdot 2\pi \cdot i)}{N}}\right| \qquad (18.24)$$

VCO Noise

$$VCO\_Noise(f) = VCO10kHz - 20 \cdot \log\left|\frac{f}{10kHz}\right| - 20 \cdot \log\left|1 + \frac{G(f \cdot 2\pi \cdot i)}{N}\right| \qquad (18.25)$$

TCXO Noise

$$TCXO\_Noise(f) = TCXO10kHz - 20 \cdot \log\left|\frac{f}{10kHz}\right| - 20 \cdot \log\left|\frac{TCXO}{Fcomp}\right| \quad (18.26)$$

$$+ 20 \cdot \log\left|\frac{G(f \cdot 2\pi \cdot i)}{1 + \frac{G(f \cdot 2\pi \cdot i)}{N}}\right|$$

Resistor Noises

Symbol	Description	Value	Units
k	Boltzman's Constant	1.3807	J/K
T	Ambient Temperature	300	K
VnR2	Noise Voltage Generated by Resistor R2	4.0704	$nV/\sqrt{Hz}$
VnR3	Noise Voltage Generated by Resistor R3	10.0614	$nV/\sqrt{Hz}$
VnR4	Noise Voltage Generated by Resistor R4	23.3382	$nV/\sqrt{Hz}$

Calculate Thermal Noise Before Transfer Functions are Applied

$$R\_Noise(R) = \sqrt{4 \cdot T \cdot k \cdot R} \quad (18.27)$$

$$VnR2 = R\_Noise(R2) \quad (18.28)$$
$$VnR3 = R\_Noise(R3)$$
$$VnR4 = R\_Noise(R4)$$

Find Resistor Noise for R2

$$Z1\_R2(s) = \frac{1}{s \cdot C2} + R2 \quad (18.29)$$

$$Z2\_R2(s) = R3 + \frac{1 + s \cdot C4 \cdot R4}{s \cdot (C3 + C4) + s^2 \cdot C3 \cdot C4 \cdot R4} \quad (18.30)$$

$$Z\_R2(s) = \frac{Z2\_R2(s)}{1 + s \cdot C1 \cdot Z2\_R2(s)} \tag{18.31}$$

$$Z3\_R2(s) = \frac{1}{1 + s \cdot (C3 \cdot R3 + C4 \cdot R4 + C4 \cdot R3) + s^2 \cdot C3 \cdot C4 \cdot R3 \cdot R4} \tag{18.32}$$

$$T\_R2(s) = \frac{1}{1 + \frac{G(s)}{N}} \cdot \frac{Z\_R2(s)}{Z1\_R2(s) + Z\_R2(s)} \cdot Z3\_R2(s) \tag{18.33}$$

$$R2\_Noise(f) = 20 \cdot \log \left| \frac{\sqrt{2} \cdot VnR2 \cdot T\_R2(2\pi \cdot f \cdot i) \cdot Kvco}{2 \cdot f} \right| \tag{18.34}$$

Find Resistor Noise for **R3**

$$Z1\_R3(s) = \frac{1 + s \cdot C2 \cdot R2}{s \cdot (C1 + C2) + s^2 \cdot C1 \cdot C2 \cdot R2} + R3 \tag{18.35}$$

$$Z2\_R3(s) = R3 + \frac{1 + s \cdot C4 \cdot R4}{s \cdot (C3 + C4) + s^2 \cdot C3 \cdot C4 \cdot R4} \tag{18.36}$$

$$T\_R3(s) = \frac{1}{1 + \frac{G(s)}{N}} \cdot \frac{Z3\_R3(s)}{Z1\_R3(s) + Z3\_R3(s)} \cdot \frac{1}{1 + s \cdot C4 \cdot R4} \tag{18.37}$$

$$R3\_Noise(f) = 20 \cdot \log \left| \frac{\sqrt{2} \cdot Vn\_R3 \cdot T\_R3(2\pi \cdot f \cdot i) \cdot Kvco}{2 \cdot f} \right| \tag{18.38}$$

Find Resistor Noise for R4

$$Z1\_R4(s) = \frac{1 + s \cdot C2 \cdot R2}{s \cdot (C1 + C2) + s^2 \cdot C1 \cdot C2 \cdot R2} + R3 \tag{18.39}$$

$$Z2\_R4(s) = R4 + \frac{R3 + Z1\_R4(s)}{1 + s \cdot C3 \cdot R3 + s \cdot C3 \cdot Z1\_R4(s)} \tag{18.40}$$

$$T\_R4(s) = \frac{1}{1 + \frac{G(s)}{N}} \cdot \frac{1}{1 + s \cdot C4 \cdot Z2\_R4(s)} \tag{18.41}$$

$$R4\_Noise(f) = 20 \cdot \log \left| \frac{\sqrt{2} \cdot Vn\_R4 \cdot T\_R4(2\pi \cdot f \cdot i) \cdot Kvco}{2 \cdot f} \right| \tag{18.42}$$

Calculate Total Noise

$$Total\_Noise(f) = 10 \cdot \log \left| \begin{array}{l} 10^{PLL\_Noise(f)/10} + 10^{VCO\_Noise(f)/10} + 10^{TCXO\_Noise(f)/10} \\ + 10^{R2\_Noise(f)/10} + 10^{R3\_Noise(f)/10} + 10^{R4\_Noise(f)/10} \end{array} \right| \tag{18.43}$$

Calculate RMS Phase Error, Error Vector Magnitude, and Jitter

$$RMS\_Phase\_Error = \frac{180}{\pi} \cdot \sqrt{2 \cdot \int_{1.7\,kHz}^{200\,kHz} 10^{Total\_Noise(f)/10} df} \tag{18.44}$$

$$EVM = \frac{\pi}{180} \cdot RMS\_Phase\_Error \tag{18.45}$$

$$Jitter = \frac{1}{Fout} \cdot \frac{RMS\_Phase\_Error}{360°} \tag{18.46}$$

Calculate RMS Frequency Error

$$RMS\_Frequency\_Error = \sqrt{2 \cdot \int_{1.7\,kHz}^{200\,kHz} 10^{Total\_Noise(f)/10} \cdot f^2 df} \tag{18.47}$$

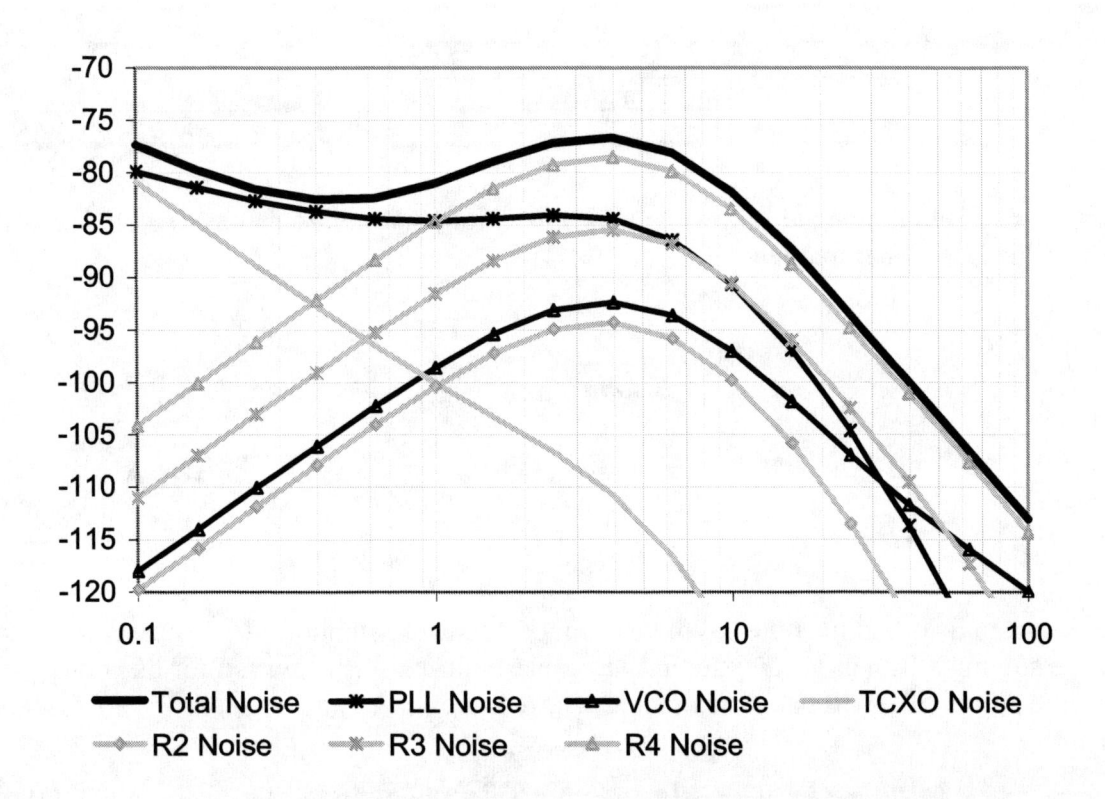

Symbol	Description	Value	Units
Total_Noise(1 kHz)	Phase Noise at 1 kHz offset	-80.16	dBc/Hz
Total_Noise(10 kHz)	Phase Noise at 10 kHz offset	-82.01	dBc/Hz
Total_Noise(100 kHz)	Phase Noise at 100 kHz offset	-113.75	dBc/Hz
RMS_Phase_Error	Root Mean Square Phase Error	1.0182	deg
EVM	Error Vector Magnitude	1.7772	%
Jitter	Jitter	5.4852	pS
RMS_Frequency_Error	Root Mean Square Frequency Error	202.0141	Hz
EVM	Error Vector Magnitude	1.7772	%

Lock Time Analysis

Symbol	Description	Value	Units
f2	Final Frequency	915	MHz
f1	Starting Frequency	889	MHz
tol	Settling tolerance within which PLL is considered locked	1000	Hz

The first step is to define the following constants. Note that the frequency used to calculate the N value is the final frequency.

$$N = \frac{f2}{Fcomp} \tag{18.48}$$

$$K = \frac{K\phi \cdot Kvco}{N} \tag{18.49}$$

The next step is to find the poles of the closed loop transfer function. In this case, it would involve solving a fifth order polynomial. Because a closed form solution for this does not exist, the polynomial will be approximated with a fourth order polynomial.

$$A2 \cdot p^4 + A1 \cdot p^3 + A0 \cdot p^2 + K \cdot T2 \cdot p + K = 0 \tag{18.50}$$

Symbol	Description	Value	Units
N	N Value for final settling frequency	4575	MHz
$K \cdot T2$	Constant	3.2787×10^{-3}	$\frac{1}{\Omega}$
K	Constant	32.7869	$\frac{1}{s\Omega}$
p_0	Closed Loop Transfer Function Pole	-1.7778×10^5	Hz
p_1	Closed Loop Transfer Function Pole	-5.6842×10^4	Hz
p_2	Closed Loop Transfer Function Pole	-3.6539×10^4	Hz
p_3	Closed Loop Transfer Function Pole	-2.0237×10^4	Hz

Calculate the Closed Loop Transfer Function Constants

$$B_0 = \frac{K \cdot (f2 - f1)}{A2} \cdot \frac{1}{(p_0 - p_1) \cdot (p_0 - p_2) \cdot (p_0 - p_3)} \tag{18.51}$$

$$B_1 = \frac{K \cdot (f2 - f1)}{A2} \cdot \frac{1}{(p_1 - p_0) \cdot (p_1 - p_2) \cdot (p_1 - p_3)} \tag{18.52}$$

$$B_2 = \frac{K \bullet (f2 - f1)}{A2} \bullet \frac{1}{(p_2 - p_0) \bullet (p_2 - p_1) \bullet (p_2 - p_3)} \qquad (18.53)$$

$$B_3 = \frac{K \bullet (f2 - f1)}{A2} \bullet \frac{1}{(p_3 - p_0) \bullet (p_3 - p_1) \bullet (p_3 - p_2)} \qquad (18.54)$$

Calculate the Transient Response and Exponential Envelope

$$F(t) = f2 + \sum_{i=0}^{3} \left[B_i \bullet \left(\frac{1}{p_i} + C2 \bullet R2 \right) \right] \qquad (18.55)$$

$$E(t) = \left| B_i \bullet \left(\frac{1}{p_i} + C2 \bullet R2 \right) \right| \qquad (18.56)$$

Symbol	Description	Value	Units
B_0	Transient Function Constant	-7.2199×10^{10}	$\frac{1}{s^2}$
B_1	Transient Function Constant	-2.1616×10^{12}	$\frac{1}{s^2}$
B_2	Transient Function Constant	-4.1560×10^{12}	$\frac{1}{s^2}$
B_3	Transient Function Constant	2.0666×10^{12}	$\frac{1}{s^2}$
Peak Time	Peak Time (extrapolated from curve)	93	μs
Lock Time	Lock Time (extrapolated from exponential envelope)	571	μs

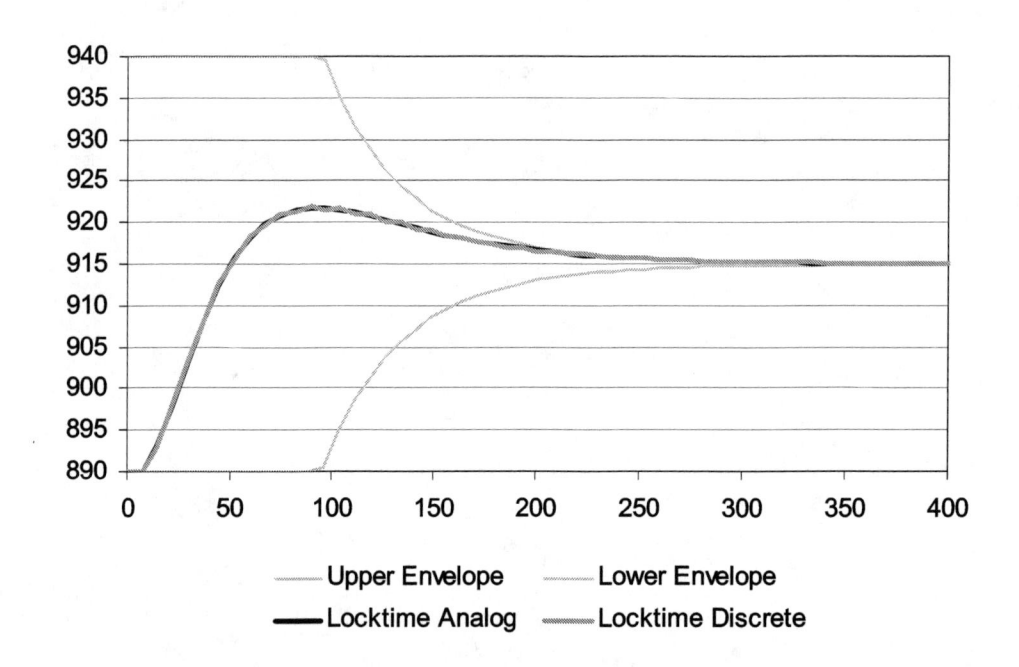

Note that although only the analog model lock time calculations are shown, the graph above also shows the curve for the discrete lock time as well.

PLL Design

Chapter 19 Fundamentals of PLL Passive Loop Filter Design

Introduction

This chapter discusses the many technical issues that come with loop filter design. Loop filter design involves choosing the proper loop filter topology, loop filter order, phase margin, loop bandwidth, and pole ratios. Once these are chosen, the poles and zero of the filter can be determined. From these, the loop filter components can then be calculated. This chapter discusses the fundamental principles that are necessary for an understanding of loop filter design.

Determining the Loop Filter Topology and Order

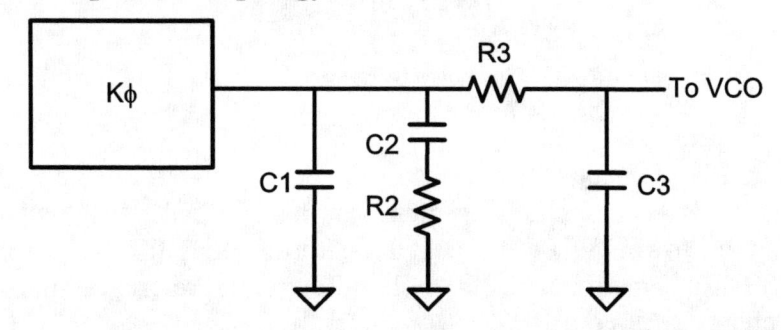

Figure 19.1 *A Third Order Passive Loop Filter*

A third order passive loop filter is show above. Passive loop filters are usually recommended above active loop filters, because adding active devices adds phase noise, complexity, and cost. However, there are cases where an active filter is necessary. The most common case arises when the maximum PLL charge pump voltage is lower than the VCO tuning voltage requirements. If higher tuning voltages are supplied to a VCO, then either the tuning range can be expanded or the phase noise reduced.

In terms of filter order, the most basic is the second order filter. Additional RC low pass filtering stages can be added to reduce the reference spurs. The impact of adding these additional stages is discussed in other chapters. In Figure 19.1 , *R3* and *C3* form an additional low pass filtering stage.

Choosing the Phase Margin, Loop Bandwidth, and Pole Ratios

The phase margin (ϕ) relates to the stability of a system. This parameter is typically chosen between 40 and 55 degrees. Simulations show that a phase margin of about 48 degrees yields the optimal lock time. Higher phase margins may decrease peaking response of the loop filter at the expense of degrading the lock time. For minimum RMS phase error designs, 50 degrees is a good starting point for phase margin.

The loop bandwidth (*Fc*) is the most critical parameter of the loop filter. Choosing the loop bandwidth too small will yield a design with improved reference spurs and RMS phase error, but all at the expense of increased lock time. Choosing the loop bandwidth too wide will result in improved lock time at the expense of increased reference spurs and RMS phase error. The suggested method of choosing the loop bandwidth is to choose it so that it is sufficient to meet the lock time requirement with sufficient margin. In cases where there is no lock time requirement, then it makes sense to choose the loop bandwidth at the frequency where the PLL noise equals the VCO noise for an optimal RMS phase error design. For a minimum reference spur design, the narrower the loop bandwidth, the lower the spurs. However, at some point the loop filter component values will be come unrealistically large.

The pole ratios (**T31**, **T41**, ..) have less impact on the design than the loop bandwidth, but still are important. They tell the ratio of each pole, relative to the pole **T1**, for instance:

$$\begin{aligned} T1 &= T11 \cdot T1 \\ T3 &= T31 \cdot T1 \\ T4 &= T41 \cdot T1 \end{aligned} \qquad (19.1)$$

(Note **T11** is trivial and always equal to 1)

It will be shown in a later chapter that choosing all pole ratios to be one is theoretically the lowest spur solution. However, choosing them smaller can make sense when the capacitor in the loop filter next to the VCO is not at least three times the VCO input capacitance (typically 10 – 100 pF). The impact of the pole ratios on the reference spurs is explained in depth in another chapter.

The Loop Filter Impedance and Open Loop Gain

The loop filter impedance is defined as the output voltage at the VCO divided by current injected at the PLL charge pump. The expression for the loop filter impedance and the corresponding poles and zeros are shown below for various filter orders.

$$Z(s) = \frac{1 + s \cdot T2}{A0 \cdot s \cdot (1 + s \cdot T1) \cdot (1 + s \cdot T3) \cdot (1 + s \cdot T4)} \qquad (19.2)$$

Parameter	Second Order Filter	Third Order Filter	Fourth Order Filter
T1	$\dfrac{R2 \cdot C2 \cdot C1}{A0}$	$\dfrac{R2 \cdot C2 \cdot C1}{A0}$ *	$\dfrac{R2 \cdot C2 \cdot C1}{A0}$ *
T2	$R2 \cdot C2$	$R2 \cdot C2$	$R2 \cdot C2$
T3	0	$R3 \cdot C3$ *	$R3 \cdot C3$ *
T4	0	0	$R4 \cdot C4$ *
A0	$C1 + C2$	$C1 + C2 + C3$	$C1 + C2 + C3 + C4$

* This indicates this formula is approximate, not exact

Table 19.1 *Impedance Parameters for Various Filter Orders*

Once the impedance ($Z(s)$), charge pump gain ($K\phi$), and VCO Gain ($Kvco$) are known, then the open loop gain ($G(s)$) is given below:

$$G(s) = \frac{K\phi \bullet Kvco}{s} \bullet Z(s) \qquad (19.3)$$

Determining the Time Constants

This method of determining the poles and zeros is taken from Application Note 1001 by National Semiconductor. The phase margin is specified as 180 degrees plus the phase of the forward loop gain, where the forward loop gain is specified as the open loop gain divided by the N divider value. Therefore, it is true that:

$$\phi = 180 + arctan(\omega c \bullet T2) - arctan(\omega c \bullet T1) - arctan(\omega c \bullet T1 \bullet T31) - arctan(\omega c \bullet T1 \bullet T41) \qquad (19.4)$$

Since ϕ and the pole ratios are known, then this can be simplified to an expression involving $T1$ and $T2$. A second expression involving $T1$ and $T2$ can be found by setting the derivative of the phase margin equal to zero at the frequency equal to the loop bandwidth. This maximizes the phase margin at this frequency. Simulations show that satisfying this condition minimizes the lock time of the PLL for second order filter.

$$\left. \frac{d\phi}{d\omega} \right|_{\omega=\omega c} = 0 = \frac{\omega c \bullet T2}{1+\omega c^2 \bullet T2^2} - \frac{\omega c \bullet T1}{1+\omega c^2 \bullet T1^2} - \frac{\omega c \bullet T1 \bullet T31}{1+\omega c^2 \bullet T1^2 \bullet T31^2} - \frac{\omega c \bullet T1 \bullet T41}{1+\omega c^2 \bullet T1^2 \bullet T41^2} \qquad (19.5)$$

Equations (19.4) and (19.5) present a system of two equations with the two unknowns, $T1$ and $T2$. The solution to these equations is presented in chapters to come. This system can always be solved numerically and in the case of a second order filter ($T3 = T4 = 0$), an elegant closed form solution exists.

Simulations show that using equation (19.5) as a constraint gives a close approximation to the loop filter with the fastest lock time, but this is not exactly correct. Using some approximations, equation (19.5) can be simplified to

$$T2 = \frac{1}{\omega c^2 \bullet (T1+T3+T4)} \qquad (19.6)$$

Since this is an approximation to a rule of thumb that is only an approximation to the exact criteria for optimal performance, it makes sense to generalize this equation as:

$$T2 = \frac{\gamma}{\omega c^2 \bullet T1 \bullet (1+T31+T43 \bullet T31)} \qquad (19.7)$$

In the above equation, γ is defined as the Gamma Optimization Factor. Now 1.0 is a good starting value for this parameter, but this parameter is discussed in depth in other chapters.

Calculating the Components from the Time Constants

Calculating the Loop Filter Coefficient A0

This is the step that is expanded in much greater detail in other chapters. However, one common concept that arises for all passive filters, regardless of the filter order, is the total capacitance. This is just the sum of all the capacitance values in the loop filter. If one considers a delta current spike, then it should be intuitive that in the long term, the voltages across all the capacitors should be the same and that its voltage would be the same as if all four capacitors values were added together. The final value theorem says this result can be found by taking the limit of $s \bullet Z(s)$ as s approaches zero. This result is $A0$, the total loop filter capacitance. $A0$ can be found by setting the forward loop gain ($G(s)$ divided by N) equal to one at the loop bandwidth.

$$A0 = \frac{K\phi \bullet Kvco}{N \bullet \omega c^2} \bullet \sqrt{\frac{(1+\omega c^2 \bullet T2^2)}{(1+\omega c^2 \bullet T1^2) \bullet (1+\omega c^2 \bullet T3^2) \bullet (1+\omega c^2 \bullet T4^2)}} \qquad (19.8)$$

Concerns with the VCO Input Capacitance

The VCO will have an input capacitance, typically on the order of 10 – 100 pF, which will add to the capacitances of the loop filter. This often becomes an issue with third and higher order loop filter designs, because the capacitor shunt with the VCO should be at least three times the VCO input capacitance to keep it from distorting the performance of the loop filter. In order to maximize this capacitance, design for the highest charge pump setting.

Concerns with Resistor Thermal Noise

The resistors in the loop filter, particularly the ones in the low pass RC filters (*R3*, *R4*, ..) generate thermal noise, which can increase the phase noise at and outside of the loop bandwidth. This starts to become a factor when these resistances are bigger than about 10 KΩ, although this is design specific. Designing for a higher charge pump current and lower pole ratios minimizes the loop filter resistors and thermal noise.

Conclusion

The equations to explicitly solve for the component values are presented in upcoming chapters, but they are all derived from these fundamental concepts and formulas presented in this chapter. The second order filter is a special case where *T3* = *T4* = *0*. The third order filter is a special case where *T3>0* and *T4 = 0*. These formulas could be easily generalized for filters of higher than fourth order, but this is more of an academic exercise than something of practical value. Note that some textbooks show a similar filter topology as presented in this chapter, except that *C1 = 0*. Although this is a stable loop filter design, this topology is not recommended, because the reference spur attenuation is not as good.

References

Keese, William O. *An Analysis and Performance Evaluation of a Passive Filter Design Technique for Charge Pump Phase-Locked Loops* Application Note 1001. National Semiconductor

Chapter 20 Equations for a Passive Second Order Loop Filter

Introduction

The second order loop filter is the least complex loop filter and allows one to explicitly solve for the component values in closed form. The second order filter has the smallest resistor thermal noise and largest capacitor next to the VCO to minimize the impact of VCO input capacitance. This filter also has maximum resistance to variations in VCO gain and charge pump gain. In cases where the first spur to be filtered is less than 10 times the loop bandwidth frequency, filter orders higher than third order do not provide much real improvement in spur levels. For the second order filter $T3 = T4 = T31 = T41 = 0$.

Loop Filter Impedance, Pole, and Zero

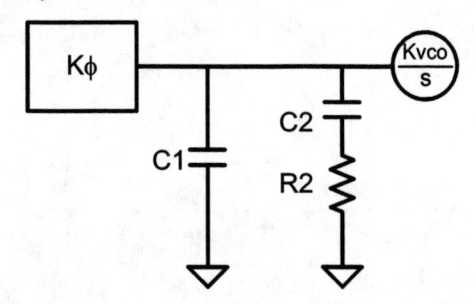

Figure 20.1 *A Second Order Passive Loop Filter*

The transfer function of a second order loop filter is given below:

$$Z(s) = \frac{1+s \cdot C2 \cdot R2}{s \cdot (C1+C2) \cdot \left(1+s \cdot \frac{C1 \cdot C2 \cdot R2}{C1+C2}\right)} = \frac{1+s \cdot T2}{s \cdot A0 \cdot (1+s \cdot T1)} \qquad (20.1)$$

From the above equation, it should be clear:

$$T2 = R2 \cdot C2 \qquad (20.2)$$
$$T1 = \frac{R2 \cdot C2 \cdot C1}{A0}$$
$$A0 = C1 + C2$$

A system of two equations and two unknowns can be established by calculating the phase margin and also setting the derivative of the phase margin equal to zero at the loop bandwidth.

$$\phi = 180 + arctan(\omega c \cdot T2) - arctan(\omega c \cdot T1) \qquad (20.3)$$

The solution to this equation is given below.

$$T2 = \frac{\gamma}{\omega c^2 \cdot T1} \qquad (20.4)$$

Substituting (20.4) into (20.3), taking the tangent of both sides, and solving yields:

$$T1 = \frac{\sqrt{(1+\gamma)^2 \cdot \tan^2 \phi + 4 \cdot \gamma} - (1+\gamma) \cdot \tan \phi}{2 \cdot \omega c} \qquad (20.5)$$

The time constant $T2$ can now be easily found using equation (20.4). The total loop filter capacitance, $A0$, can be found and $C1$ can be calculated.

$$A0 = \frac{C1 \cdot T2}{T1} = \frac{K\phi \cdot Kvco}{N \cdot \omega c^2} \cdot \sqrt{\frac{(1 + \omega c^2 \cdot T2^2)}{(1 + \omega c^2 \cdot T1^2)}} \qquad (20.6)$$

Once the total capacitance is known, the components can be easily found:

$$\Rightarrow C1 = A0 \cdot \frac{T1}{T2} \qquad (20.7)$$
$$\Rightarrow C2 = A0 - C1 \qquad (20.8)$$
$$\Rightarrow R2 = \frac{T2}{C2} \qquad (20.9)$$

Conclusion

The formulas for the second order passive loop filter have been presented in this chapter. These formulas are just a special case of the formulas presented in a previous chapter. The second order filter has an elegant solution for the component values, but higher order filters may have lower reference spurs. A particular topology of loop filter was assumed in this chapter. There is actually another topology for the second order filter that is sometimes used in active filters. For different topologies, the component values may change, but the formulas for the time constants remain the same.

Reference

Keese, William O.   *An Analysis and Performance Evaluation of a Passive Filter Design Technique for Charge Pump Phase-Locked Loops*

Appendix A: A Second Order Loop Filter Design

Design Specifications

Symbol	Description	Value	Units
Fc	Loop Bandwidth	10	kHz
ϕ	Phase Margin	49.2	degrees
γ	Gamma Optimization Parameter	1.024	none
$K\phi$	Charge Pump Gain	1	mA
$Kvco$	VCO Gain	60	MHz/V
$Fout$	Output Frequency	1960	MHz
$Fcomp$	Comparison Frequency	50	kHz

Calculate Poles and Zero

$$N = \frac{Fout}{Fcomp} \tag{20.10}$$

$$\omega c = 2 \cdot \pi \cdot Fc \tag{20.11}$$

$$T1 = \frac{\sqrt{(1+\gamma)^2 \cdot \tan^2\phi + 4\cdot\gamma} - (1+\gamma)\cdot\tan\phi}{2\cdot\omega c} \tag{20.12}$$

$$T2 = \frac{\gamma}{\omega c^2 \cdot T1} \tag{20.13}$$

Calculate Loop Filter Coefficients

$$A0 = \frac{C1 \cdot T2}{T1} = \frac{K\phi \cdot Kvco}{N \cdot \omega c^2} \cdot \sqrt{\frac{\left(1+\omega c^2 \cdot T2^2\right)}{\left(1+\omega c^2 \cdot T1^2\right)}} \tag{20.14}$$

Solve For Components

$$C1 = A0 \cdot \frac{T1}{T2} \tag{20.15}$$

$$C2 = A0 - C1 \tag{20.16}$$

$$\Rightarrow R2 = \frac{T2}{C2} \tag{20.17}$$

Results

Symbol	Description	Value	Units
N	N Counter Value	39200	none
ωc	Loop Bandwidth	6.283×10^4	rad/s
$T1$	Loop Filter Pole	5.989×10^{-6}	s
$T2$	Loop Filter Zero	4.331×10^{-5}	s
$A0$	Total Capacitance	1.052	nF
$C1$	Loop Filter Capacitor	0.145	nF
$C2$	Loop Filter Capacitor	0.906	nF
$R2$	Loop Filter Resistor	47.776	kΩ

Chapter 21 Equations for a Passive Third Order Loop Filter

Introduction

In cases where the spur to be filtered is more than ten times the loop bandwidth, a third order filter can provide some benefit. Unlike the second order loop filter, there is no closed form solution for the exact component values. Designing the loop filter involves solving for the time constants, and then determining the loop filter components from the time constants. The time constants can be calculated either by introducing approximations and writing down a closed form approximate solution, or using numerical methods to solve more precisely for the time constants. Once the time constants are found, the component values can also be calculated by introducing approximations (although the results will not be exact), or can be calculated more exactly by using numerical methods.

Note that in addition to specifying the loop bandwidth, ωc, and phase margin, ϕ, the user also has to specify the pole ratio, **T31**. This parameter can range from zero to one. A good starting value for this parameter is **0.5**.

Calculating the Loop Filter Impedance and Time Constants

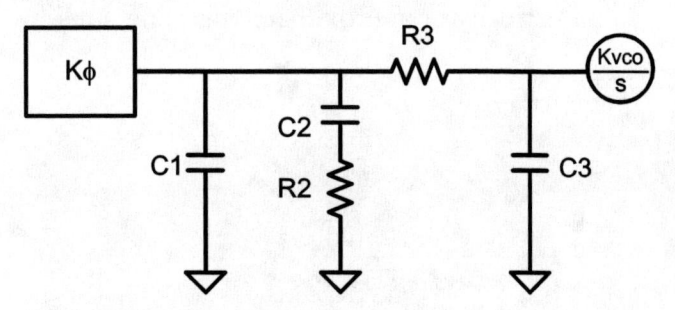

Figure 21.1 *Third Order Passive Loop Filter*

For the loop filter shown in Figure 21.1, the impedance is given below:

$$Z(s) = \frac{1+s\bullet T2}{s\bullet A0\bullet(1+s\bullet T1)\bullet(1+s\bullet T3)} = \tag{21.1}$$

$$\frac{1+s\bullet C2\bullet R2}{s\bullet(A2\bullet s^2 + A1\bullet s + A0)} \tag{21.2}$$

$$T2 = R2\bullet C2 \tag{21.3}$$

$$\begin{aligned} A2 &= A0\bullet T1\bullet T3 = C1\bullet C2\bullet R2\bullet C3\bullet R3 \\ A1 &= A0\bullet(T1+T3) = C2\bullet C3\bullet R2 + C1\bullet C2\bullet R2 + C1\bullet C3\bullet R3 + C2\bullet C3\bullet R3 \\ A0 & = C1 + C2 + C3 \end{aligned} \tag{21.4}$$

Loop Filter Calculation

Calculation of Time Constants

By setting the derivative of the phase margin equal to zero, the following relationship is obtained:

$$T2 = \frac{\gamma}{\omega c^2 \cdot T1 \cdot (1+T31)} \tag{21.5}$$

The phase margin is given by:

$$\phi = tan^{-1}(\omega c \cdot T2) - tan^{-1}(\omega c \cdot T1) - tan^{-1}(\omega c \cdot T3) \tag{21.6}$$

$$\phi = tan^{-1}\left(\frac{\gamma}{\omega c \cdot T1 \cdot (1+T31)}\right) - tan^{-1}(\omega c \cdot T1) - tan^{-1}(\omega c \cdot T1 \cdot T31) \tag{21.7}$$

This can either be solved numerically or approximately as shown below:

$$tan(x) \approx x \approx tan^{-1}(x) \tag{21.8}$$

$$T1 \approx \frac{sec(\phi) - tan(\phi)}{\omega c \cdot (1+T31)} \tag{21.9}$$

Once *T1* is known, *T2* and *T3* can be easily found:

$$T3 = T1 \cdot T31 \tag{21.10}$$

$$T2 = \frac{\gamma}{\omega c^2 \cdot (T1+T3)} \tag{21.11}$$

Solution of Component Values from Time Constants

The first step to is to calculate the total capacitance:

$$A0 = \frac{K\phi \cdot Kvco}{\omega c^2 \cdot N} \cdot \sqrt{\frac{1+\omega c^2 \cdot T2^2}{(1+\omega c^2 \cdot T1^2) \cdot (1+\omega c^2 \cdot T3^2)}} \tag{21.12}$$

True Loop Filter Impedance

The true impedance of the filter is given by:

$$Z(s) = \frac{1+s \cdot T2}{s \cdot (1+s \cdot T1) \cdot (1+s \cdot T3)} \cdot \frac{1}{A0} \qquad (21.13)$$

Recall that the loop filter components relate to the time constants in the following manner for a passive filter.

$$
\begin{aligned}
A2 &= A0 \cdot T1 \cdot T3 = C1 \cdot C2 \cdot R2 \cdot C3 \cdot R3 \\
A1 &= A0 \cdot (T1+T3) = C2 \cdot C3 \cdot R2 + C1 \cdot C2 \cdot R2 + C1 \cdot C3 \cdot R3 + C2 \cdot C3 \cdot R3 \\
A0 &= C1 + C2 + C3
\end{aligned} \qquad (21.14)
$$

Now the first step to solving for the components is to choose the component $C1$. There are many possible choices, but the optimal choice is the one that maximizes the capacitor $C3$. This is desirable because it minimizes the impact of the VCO capacitance and also resistor thermal noise due to $R3$. Although the choice of $C1$ that minimizes $R3$ is slightly different than the choice of $C1$ that maximizes $C3$, these two values are very close, and making $C3$ larger attenuates the noise due to resistor $R3$ more. The justification for the choice of $C1$ is shown in the appendix.

$$C1 = \frac{A2}{T2^2} \cdot \left(1 + \sqrt{1 + \frac{T2}{A2} \cdot (T2 \cdot A0 - A1)}\right) \qquad (21.15)$$

Combining these equations yields:

$$A1 = T2 \cdot C1 - \frac{A2 \cdot A0}{T2 \cdot C1} - \frac{A2 \cdot C3}{T2 \cdot C1} \qquad (21.16)$$

The above equation can be solved in order to express $C3$ in terms of $C1$:

$$C3 = \frac{-T2^2 \cdot C1^2 + T2 \cdot A1 \cdot C1 - A2 \cdot A0}{T2^2 \cdot C1 - A2} \qquad (21.17)$$

$C2$ and the other components can now be easily found.

$$C2 = A0 - C1 - C3 \qquad (21.18)$$

$$R2 = \frac{T2}{C2} \qquad (21.19)$$

$$R3 = \frac{A2}{C1 \bullet C3 \bullet T2} \qquad (21.20)$$

Conclusion

This chapter has presented a method for calculating a third order passive loop filter. Unlike the second order filter equations, there is no closed form solution for the time constants, although it is easy to solve for them numerically. Once these time constants are known, then the component values can be calculated. For those who wish to avoid these numerical methods, simplified approximate equations for the time constants have also been presented.

Regardless of the filter calculation method used, the VCO input capacitance adds to capacitor $C3$, so this component should be at least four times the VCO input capacitance. In many circumstances, this is not possible. If the value of $T31$ is decreased, then the capacitor $C3$ will become larger and the resistor $R3$ will become smaller. Choosing $C3$ as large as possible also corresponds to choosing $R3$ as small as possible. It is desirable to not have the $R3$ resistor too large, or else the thermal noise from this resistor can add to the out of band phase noise.

References

Keese, William O. *An Analysis and Performance Evaluation of a Passive Filter Design Technique for Charge Pump Phase-Locked Loops*

Appendix A: A Third Order Loop Filter Design

Design Specifications

Symbol	Description	Value	Units
Fc	Loop Bandwidth	2	kHz
ϕ	Phase Margin	47.1	degrees
γ	Gamma Optimization Parameter	1.136	none
$K\phi$	Charge Pump Gain	4	mA
$Kvco$	VCO Gain	30	MHz/V
$Fout$	Output Frequency	1392	MHz
$Fcomp$	Comparison Frequency	60	kHz
$T31$	Ratio of pole $T3$ to Pole $T1$	0.6	none

Calculate Poles and Zero

$$N = \frac{Fout}{Fcomp} \quad (21.21)$$

$$\omega c = 2 \bullet \pi \bullet Fc \quad (21.22)$$

T1 is the only unknown. Use the Exact Method to Solve for T1 Using Numerical Methods

$$\phi = tan^{-1}\left(\frac{\gamma}{\omega c \bullet T1 \bullet (1+T31)}\right) - tan^{-1}(\omega c \bullet T1) - tan^{-1}(\omega c \bullet T1 \bullet T31) \quad (21.23)$$

$$T3 = T1 \bullet T31 \quad (21.24)$$

$$T2 = \frac{\gamma}{\omega c^2 \bullet (T1+T3)} \quad (21.25)$$

Calculate Loop Filter Coefficients

$$A0 = \frac{K\phi \bullet Kvco}{\omega c^2 \bullet N} \bullet \sqrt{\frac{1+\omega c^2 \bullet T2^2}{(1+\omega c^2 \bullet T1^2) \bullet (1+\omega c^2 \bullet T3^2)}} \quad (21.26)$$

$$A1 = A0 \bullet (T1+T3) \quad (21.27)$$

$$A2 = A0 \bullet T1 \bullet T3 \quad (21.28)$$

Symbol	Description	Value	Units
N	N Counter Value	23200	none
ω_c	Loop Bandwidth	1.2566×10^4	rad/s
T1	Loop Filter Pole	2.0333×10^{-5}	s
T2	Loop Filter Zero	2.2112×10^{-4}	s
T3	Loop Filter Zero	1.2200×10^{-5}	s
A0	Total Capacitance	92.6372	nF
A1	First order loop filter coefficient	3.0138×10^{-3}	nFs
A2	Second Order loop filter coefficient	2.2980×10^{-8}	nFs^2

Solve For Components

$$C1 = \frac{A2}{T2^2} \bullet \left(1 + \sqrt{1 + \frac{T2}{A2} \bullet (T2 \bullet A0 - A1)} \right) \qquad (21.29)$$

$$C3 = \frac{-T2^2 \bullet C1^2 + T2 \bullet A1 \bullet C1 - A2 \bullet A0}{T2^2 \bullet C1 - A2} \qquad (21.30)$$

$$C2 = A0 - C1 - C3 \qquad (21.31)$$

$$R2 = \frac{T2}{C2} \qquad (21.32)$$

$$R3 = \frac{A2}{C1 \bullet C3 \bullet T2} \qquad (21.33)$$

Results

Symbol	Description	Value	Units
C1	Loop Filter Capacitor	6.5817	nF
C2	Loop Filter Capacitor	85.5896	nF
C3	Loop Filter Capacitor	0.4660	nF
R2	Loop Filter Resistor	2.5835	kΩ
R3	Loop Filter Resistor	33.8818	kΩ

Appendix B: Optimal Choice of *C1* and Verification it Leads to Positive Components

The process to fully justify the choice of *C1* to express *C3* as a function of *C1* and apply the first derivative to find the critical points. Then it will be proven that the largest critical point is indeed the value of *C1* that yields the largest value for *C3*, provided that *C1*>0. Then it will be shown that the values attained for *C2* and *C3* from this optimal choice of *C1* are always positive. Once all the capacitor values are known to be positive, one can easily show the resistor values must be positive as well. As a residual result of these calculations it is also shown that *T31* must be strictly less than one.

Find the Critical Points for the Expression for C3 in terms of C1 and Find the Critical Point

The first step is to apply the first derivative to equation (21.17) and equate this to zero in order to find the critical points.

$$\frac{dC3}{dC1} = -\frac{C1^2 - \left(\frac{2 \bullet A2}{T2^2}\right) \bullet C1 + \left(\frac{A1 \bullet A2}{T2^3} - \frac{A2 \bullet A0}{T2^2}\right)}{\left(C1 - \frac{A2}{T2^2}\right)^2} = 0 \quad (21.34)$$

By setting the numerator equal to zero and solving, the following result is obtained.

$$C1 = \frac{A2}{T2^2} \bullet \left(1 \pm \sqrt{1 + \frac{T2}{A2} \bullet (T2 \bullet A0 - A1)}\right) \quad (21.35)$$

Determine Which Critical Point is the Correct One and Verify that it is a Global Maximum for C1>0

Recall that if the second derivative is negative, it indicates the critical point is a local maximum, and if it is positive, it indicates that it is a local minimum. Taking another derivative of (21.17) yields:

$$\frac{d^2 C3}{dC1^2} = \frac{-2}{C1 - \frac{A2}{T2^2}} \quad (21.36)$$

Now if one uses the critical point in equation (21.35) with the negative sign in front of the square root sign, this will be a local minimum, since the derivative would be positive. It also follows that the using the critical point with the positive root yields a local maximum. Now within a small neighborhood of this largest critical point, for *C1* larger than this value, the value of *C3* is decreasing. Since there are no critical points larger than this value, it follows that there can be no global maximum value for *C1* larger than the largest critical point. Now for an infinitesimally small neighborhood around this critical point, but for slightly smaller values for $C1$, the slope is negative, and it does not become positive again until one reaches the smaller critical point. But since this critical point can be shown to be less than 0, it follows that the following value yields the largest possible value for $C3$, provided that $C1>0$.

$$C1 = \frac{A0 \cdot T1 \cdot T3}{T2^2} \cdot \left(1 + \sqrt{1 + \frac{T2}{T1 \cdot T3} \cdot (T2 - T1 - T3)}\right) \tag{21.37}$$

Find the Restrictions on $C1$ to Ensure That $C3$ is Positive

Now the expression for $C3$ can be expressed in terms of time constants, $C1$, and $A0$.

$$C3 = \frac{-T2^2 \cdot C1^2 + T2 \cdot A0 \cdot (T1 + T3) \cdot C1 - T1 \cdot T3 \cdot A0^2}{T2^2 \cdot C1 - T1 \cdot T3 \cdot A0} \tag{21.38}$$

Now it is easy to see by inspection that the denominator will be positive for the optimal choice of $C1$, but the numerator is not so obvious. Using the quadratic formula and simplifying yields the restrictions on $C1$ which are necessary to make $C3>0$. It is assumed, by definition, that $T3>T1$.

$$\frac{T3}{T2} \cdot A0 < C1 < \frac{T1}{T2} \cdot A0 \tag{21.39}$$

Now applying these above restrictions to the value of $C1$ shows that they will be satisfied, provided the following conditions are met.

$$T2 > T1 > T3 \tag{21.40}$$

Note that $T2>T1$ is required for stability and $T1>T3$ is true, since $T1$ is defined as the larger of the two time constants. Note this also shows that if one chooses $T1 = T3$, the capacitor $C3$ will be zero, which is indeed the case. Therefore there is an additional requirement implied by this.

Find Restrictions on $C1$ to Ensure that $C2$ is Positive

Applying (21.39) and seeking the condition that ensures that $C2$ is positive is yields the following constraint that is always satisfied for a stable loop filter, since $T2>T1+T3$.

$$C1 + C3 < A0 \tag{21.41}$$
$$\Rightarrow T1 \cdot T3 + T2 \cdot (T2 - T1 - T3)$$

Chapter 22 Equations for a Passive Fourth Order Loop Filter

Introduction

In cases where the spur to be filtered is more than twenty times the loop bandwidth, a fourth order filter can provide some benefit. The fourth order filter also becomes necessary with some delta sigma PLLs.

Many challenges come with the 4th order loop filter design. The challenge with the fourth order filter is definitely solving for the components. Unlike lower order filters, it is actually possible to design a stable fourth order loop filter that has all real components, yet has complex poles. Although this may prove advantageous, this chapter assumes all poles in the filter are real. It is also a challenge to have any idea if the component values yield the maximum possible value for $C4$. In the case of a 3rd order loop filter, it was possible to solve for components and prove that the solution always yielded the maximum possible capacitor next to the VCO. In the case of a 4th order loop filter, this is possible, but because the routine is so complicated and has so many problems with always converging to a solution with all positive component values, it makes sense to introduce simplifications. The basic strategy presented in this chapter is to design a third order loop filter and then perturb this into a fourth order loop filter. The solution yields exactly the parameters designed for, but does not yield the maximum possible value for $C4$, although it comes very close.

Calculating the Loop Filter Impedance and Time Constants

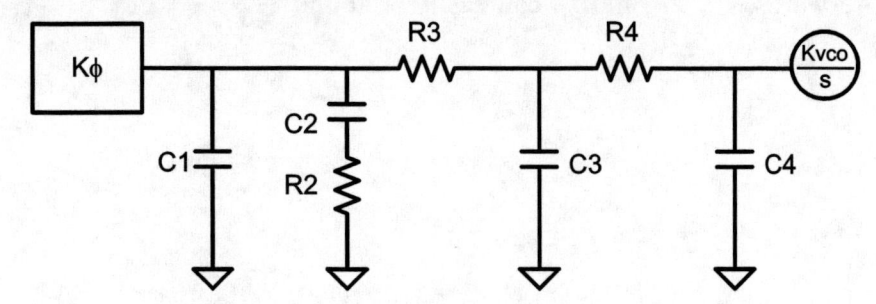

Figure 22.1 *Fourth Order Passive Loop Filter*

For the loop filter shown in Figure 22.1, the impedance is given below:

$$Z(s) = \frac{1+s \bullet T2}{s \bullet A0 \bullet (1+s \bullet T1) \bullet (1+s \bullet T3) \bullet (1+s \bullet T4)} = \frac{1+s \bullet C2 \bullet R2}{s \bullet (A3 \bullet s^3 + A2 \bullet s^2 + A1 \bullet s + A0)} \qquad (22.1)$$

$$A3 = R2 \cdot R3 \cdot R4 \cdot C1 \cdot C2 \cdot C3 \cdot C4 \qquad (22.2)$$

$$A2 = C1 \cdot C2 \cdot R2 \cdot R3 \cdot (C3+C4) + C3 \cdot C4 \cdot R3 \cdot R4 \cdot (C1+C2)$$
$$+ C1 \cdot C2 \cdot R2 \cdot C4 \cdot R4$$

$$A1 = R2 \cdot C2 \cdot (C1+C3+C4) + R3 \cdot (C1+C2) \cdot (C3+C4)$$
$$+ R4 \cdot C4 \cdot (C1+C2+C3)$$

$$A0 = C1+C2+C3+C4$$

Loop Filter Calculation

Calculation of Time Constants

The phase margin is given by:

$$\phi = tan^{-1}(\omega c \cdot T2) - tan^{-1}(\omega c \cdot T1) - tan^{-1}(\omega c \cdot T3) - tan^{-1}(\omega c \cdot T4) \qquad (22.3)$$

$$\phi = tan^{-1}\left(\frac{\gamma}{\omega c \cdot T1 \cdot (1+T31)}\right) - tan^{-1}(\omega c \cdot T1) \qquad (22.4)$$
$$- tan^{-1}(\omega c \cdot T1 \cdot T31) - tan^{-1}(\omega c \cdot T1 \cdot T31 \cdot T43)$$

Now $T1$ is the only unknown in the equation above, and this can be solved for numerically for $T1$, and afterwards, $T2$, $T3$, and $T4$ can easily be found.

$$T3 = T1 \cdot T31 \qquad (22.5)$$

$$T4 = T1 \cdot T31 \cdot T43 \qquad (22.6)$$

$$T2 = \frac{\gamma}{\omega c^2 \cdot (T1+T3+T4)} \qquad (22.7)$$

Solution of Component Values from Time Constants

Calculation of Filter Impedance Coefficients

The loop filter coefficients can be calculated as follows:

$$A0 = \frac{K\phi \cdot Kvco}{\omega c^2 \cdot N} \cdot \sqrt{\frac{1+\omega c^2 \cdot T2^2}{(1+\omega c^2 \cdot T1^2) \cdot (1+\omega c^2 \cdot T3^2) \cdot (1+\omega c^2 \cdot T4^2)}} \quad (22.8)$$

$$A1 = A0 \cdot (T1+T3+T4) \quad (22.9)$$

$$A2 = A0 \cdot (T1 \cdot T3 + T1 \cdot T4 + T3 \cdot T4) \quad (22.10)$$

$$A3 = A0 \cdot T1 \cdot T3 \cdot T4 \quad (22.11)$$

Relation of Filter Impedance Coefficients to Component Values

Relating the filter impedance coefficients and zero, $T2$, to the component values yields a system of 5 equations and seven unknowns. The unknowns are the components $C1$, $C2$, $C3$, $C4$, $R2$, $R3$, and $R4$.

$$\begin{aligned}
T2 &= R2 \cdot C2 \\
A3 &= R2 \cdot R3 \cdot R4 \cdot C1 \cdot C2 \cdot C3 \cdot C4 \\
A2 &= R2 \cdot R3 \cdot C1 \cdot C2 \cdot (C3+C4) + R4 \cdot C4 \cdot (C2 \cdot C3 \cdot R3 + C1 \cdot C3 \cdot R3 + C1 \cdot C2 \cdot R2) \\
A1 &= R2 \cdot C2 \cdot (C1+C3+C4) + R3 \cdot (C1+C2) \cdot (C3+C4) + R4 \cdot C4 \cdot (C1+C2+C3) \\
A0 &= C1+C2+C3+C4
\end{aligned} \quad (22.12)$$

Since there are only five equations and seven unknowns, one can actually choose two parameters. However, it is not nearly that simple, since it is not at all obvious which components should be chosen, how to choose them to maximize $C4$, and how to choose them in such a way that the other five components values will be real and positive. These above three issues have been explored in depth, although the reader will be spared all of this. It turns out that these equations can be solved exactly. In addition to this, there does exist a method that generates the largest possible value for $C4$. The problem with this method is that it is hideously complicated, requires numerical methods to solve, and can lead to negative component values for the other components. The method presented here does not exactly maximize $C4$, but is orders of magnitude easier to use, and for all test cases, yields positive component values. It also turns out that it gets pretty close to maximum value for $C4$. Although many different avenues have been explored, the most robust solution seems to be to choose $C1$ and $R3$ first, and then solve for the other component values.

Choosing the Components C1 and R3

Recall that for the third order loop filter, it was shown that the solution always yielded positive components and yielded the maximum capacitor for **C3**. The concept here is choose **C1** and **R3** from the third order loop filter design, and then find the other components. The solution to the third order loop filter is as follows:

$$a0 = A0 = C1 + C2 + C3 + C4 \qquad (22.13)$$
$$a1 = a0 \bullet (T1 + T3)$$
$$a2 = a0 \bullet T1 \bullet T3$$

$$C1 = \frac{a2}{T2^2} \bullet \left(1 + \sqrt{1 + \frac{T2}{a2} \bullet (T2 \bullet a0 - a1)}\right) \qquad (22.14)$$

$$c3 = \frac{-T2^2 \bullet C1^2 + T2 \bullet a1 \bullet C1 - a2 \bullet a0}{T2^2 \bullet C1 - a2} \qquad (22.15)$$

$$R3 = \frac{a2}{C1 \bullet c3 \bullet T2} \qquad (22.16)$$

Note that in the above equations, *a0*, *a1*, *a2*, and *c3* are intentionally not capitalized due to the fact that they are only intermediate calculations for these values, and not the actual loop filter impedance parameters or capacitance.

Determination of C2

Now if **C1** and **C2** were specified first, then the equations would be greatly simplified. However, the problem is that the loop filter values are extremely sensitive to the choice of **C2**. Once **C1** is specified, then if **C2** is not chosen in a very narrow range, then negative component the solution will yield negative component values for some of the other components. However, if **C1** and **R3** are chosen first, this problem does not seem to exhibit itself.

The equations simplified can be rewritten in this form:

$$\frac{A3}{T2 \bullet R3 \bullet C1} = R4 \bullet C4 \bullet C3 \qquad (22.17)$$

$$\frac{A2}{A3} - \frac{1}{T2} = \frac{1}{C4 \bullet R4} + \frac{1}{C3 \bullet R4} + \frac{1}{C1 \bullet R2} + \frac{1}{C3 \bullet R3}$$

$$A1 - T2 \bullet A0 - \frac{A3}{T2 \bullet R3 \bullet C1} = -T2 \bullet C2 + R3 \bullet (C1 + C2) \bullet (C3 + C4) + R4 \bullet C4 \bullet (C1 + C2)$$

$$A0 - C1 = C2 + C3 + C4$$

Using the first equation to eliminate $R4 \cdot C4$ and rearranging yields:

$$k0 = C2 \cdot \left(\frac{1}{T2 \cdot C1} - \frac{T2 \cdot R3 \cdot C1}{A3} \right) + \frac{1}{C3 \cdot R3} \quad (22.18)$$

$$k1 = -T2 \cdot C2 + C2 \cdot [R3 \cdot (A0 - C1)] + \frac{1}{C3 \cdot R3} \cdot (C1 + C2) \cdot \left(\frac{A3}{T2 \cdot R3 \cdot C1} \right)$$

where

$$k0 = \frac{A2}{A3} - \frac{1}{T2} - \frac{(A0 - C1) \cdot T2 \cdot R3 \cdot C1}{A3}$$

$$k1 = A1 - T2 \cdot A0 - \frac{A3}{T2 \cdot R3 \cdot C1} - (A0 - C1) \cdot R3 \cdot C1$$

By combining the above equations to eliminate $\frac{1}{C3 \cdot R3}$, $C2$ can be found by solving the following quadratic equation that results:

$$a \cdot C2^2 + b \cdot C2 + c = 0 \quad (22.19)$$

where

$$a = \frac{A3}{(T2 \cdot C1)^2}$$

$$b = T2 + R3 \cdot (C1 - A0) + \frac{A3}{T2 \cdot C1} \cdot \left(\frac{1}{T2} - k0 \right)$$

$$c = k1 - \frac{k0 \cdot A3}{T2}$$

$$C2 = \frac{-b + \sqrt{b^2 - 4 \cdot a \cdot c}}{2 \cdot a}$$

Solution for Other Components

Once $C2$ is known, the other components can easily be found.

$$C3 = \frac{T2 \cdot A3 \cdot C1}{R3 \cdot [k0 \cdot T2 \cdot A3 \cdot C1 - C2 \cdot (A3 - R3 \cdot (T2 \cdot C1)^2)]} \quad (22.20)$$

$$C4 = A0 - C1 - C2 - C3$$

$$R2 = \frac{T2}{C2}$$

$$R4 = \frac{A3}{T2 \cdot R3 \cdot C1 \cdot C3 \cdot C4}$$

Conclusion

This chapter has discussed the design of a fourth order passive filter. A lot of the complexity comes in from solving for the component values, once the filter coefficients are known. Unlike the third order solution, there is no proof that the component values yielded are always positive, nor is there a proof that the capacitor next to the VCO is the largest possible, in fact, it is not. However, provided that the restriction:

$$T31 + T43 \leq 1 \qquad (22.21)$$

is followed, then there have been no cases found where the techniques presented in this chapter do not yield a realizable solution. In addition to this, the solution method presented in this chapter was compared against the solution that does yield the maximum value for the capacitor, *C4*, and the values were close. The reason that the other method was not presented is that it is much more complicated and it has problems converging to real component values in all cases. In the cases tested that it does converge to a solution with real component values, the value for the capacitor, *C4*, was only marginally smaller.

References

Keese, William O.   *An Analysis and Performance Evaluation of a Passive Filter Design Technique for Charge Pump Phase-Locked Loops*

Appendix A: A Fourth Order Loop Filter Design

Design Specifications

Symbol	Description	Value	Units
Fc	Loop Bandwidth	10	kHz
ϕ	Phase Margin	47.8	degrees
γ	Gamma Optimization Parameter	1.115	none
$K\phi$	Charge Pump Gain	4	mA
$Kvco$	VCO Gain	20	MHz/V
$Fout$	Output Frequency	900	MHz
$Fcomp$	Comparison Frequency	200	kHz
$T31$	Ratio of pole $T3$ to Pole $T1$	0.4	none
$T43$	Ratio of pole $T4$ to Pole $T1$	0.4	none

Calculate Poles and Zero

$$N = \frac{Fout}{Fcomp} \tag{22.22}$$

$$\omega c = 2 \bullet \pi \bullet Fc \tag{22.23}$$

T1 is the only unknown. Use the Exact Method to Solve for **T1** Using Numerical Methods

$$\phi = tan^{-1}\left(\frac{\gamma}{\omega c \bullet T1 \bullet (1+T31)}\right) - tan^{-1}(\omega c \bullet T1) \\ - tan^{-1}(\omega c \bullet T1 \bullet T31) - tan^{-1}(\omega c \bullet T1 \bullet T31 \bullet T43) \tag{22.24}$$

$$T3 = T1 \bullet T31 \tag{22.25}$$

$$T4 = T1 \bullet T31 \bullet T43 \tag{22.26}$$

$$T2 = \frac{\gamma}{\omega c^2 \bullet (T1+T3+T4)} \tag{22.27}$$

Calculate Intermediate Loop Filter Coefficients

$$a0 = \frac{K\phi \cdot Kvco}{\omega c^2 \cdot N} \cdot \sqrt{\frac{1+\omega c^2 \cdot T2^2}{(1+\omega c^2 \cdot T1^2) \cdot (1+\omega c^2 \cdot T3^2) \cdot (1+\omega c^2 \cdot T4^2)}} \qquad (22.28)$$

$$a1 = a0 \cdot (T1+T3) \qquad (22.29)$$

$$a2 = a0 \cdot T1 \cdot T3 \qquad (22.30)$$

$$a3 = 0 \qquad (22.31)$$

Symbol	Description	Value	Units
N	N Counter Value	4500	none
ωc	Loop Bandwidth	6.2832×10^4	rad/s
$T1$	Loop Filter Pole	4.0685×10^{-6}	s
$T2$	Loop Filter Zero	4.4500×10^{-5}	s
$T3$	Loop Filter Pole	1.6274×10^{-6}	s
$T4$	Loop Filter Pole	6.5096×10^{-7}	s
$a0$	Intermediate Total Capacitance	12.8773	nF
$a1$	Intermediate First order loop filter coefficient	7.3348×10^{-8}	nFs
$a2$	Intermediate Second Order loop filter coefficient	8.5262×10^{-11}	nFs^2
$a3$	Intermediate Third Order loop filter coefficient	0	nFs^3

Solve For Components $C1$ and $R3$

$$C1 = \frac{a2}{T2^2} \cdot \left(1 + \sqrt{1 + \frac{T2}{a2} \cdot (T2 \cdot a0 - a1)}\right) \qquad (22.32)$$

$$c3 = \frac{-T2^2 \cdot C1^2 + T2 \cdot a1 \cdot C1 - a2 \cdot a0}{T2^2 \cdot C1 - a2} \qquad (22.33)$$

$$R3 = \frac{a2}{C1 \cdot c3 \cdot T2} \qquad (22.34)$$

Symbol	Description	Value	Units
$C1$	Loop Filter Capacitor	0.7397	nF
$R3$	Loop Filter Resistor	15.3409	kΩ

Solve For *C2*

$$A0 = a0 \tag{22.35}$$

$$A1 = A0 \cdot (T1 + T3 + T4) \tag{22.36}$$

$$A2 = A0 \cdot (T1 \cdot T3 + T1 \cdot T4 + T3 \cdot T4) \tag{22.37}$$

$$A3 = A0 \cdot T1 \cdot T3 \cdot T4 \tag{22.38}$$

$$k0 = \frac{A2}{A3} - \frac{1}{T2} - \frac{(A0 - C1) \cdot T2 \cdot R3 \cdot C1}{A3} \tag{22.39}$$

$$k1 = A1 - T2 \cdot A0 - \frac{A3}{T2 \cdot R3 \cdot C1} - (A0 - C1) \cdot R3 \cdot C1$$

$$a = \frac{A3}{(T2 \cdot C1)^2} \tag{22.40}$$

$$b = T2 + R3 \cdot (C1 - A0) + \frac{A3}{T2 \cdot C1} \cdot \left(\frac{1}{T2} - k0\right)$$

$$c = k1 - \frac{k0 \cdot A3}{T2}$$

$$C2 = \frac{-b + \sqrt{b^2 - 4 \cdot a \cdot c}}{2 \cdot a} \tag{22.41}$$

Symbol	Description	Value	Units
$A0$	Total Capacitance	12.8773	nF
$A1$	First order loop filter coefficient	8.1731×10^{-5}	nFs
$A2$	Second Order loop filter coefficient	1.3301×10^{-10}	nFs^2
$A3$	Third Order loop filter coefficient	5.5502×10^{-17}	nFs^3
$k0$	Intermediate Calculation for finding $C2$	-1.0806×10^{8}	1/s
$k1$	Intermediate Calculation for finding $C2$	-6.2915×10^{-4}	nFs
a	Intermediate Calculation for finding $C2$	5.1223×10^{-8}	s/nF
b	Intermediate Calculation for finding $C2$	4.0535×10^{-5}	s
c	Intermediate Calculation for finding $C2$	-4.9437×10^{-4}	nFs
$C2$	Loop Filter Capacitor	12.0139	nF

Solve for the Other Components

$$C3 = \frac{T2 \cdot A3 \cdot C1}{R3 \cdot \left[k0 \cdot T2 \cdot A3 \cdot C1 - C2 \cdot \left(A3 - R3 \cdot (T2 \cdot C1)^2\right)\right]} \quad (22.42)$$

$$C4 = A0 - C1 - C2 - C3$$

$$R2 = \frac{T2}{C2}$$

$$R4 = \frac{A3}{T2 \cdot R3 \cdot C1 \cdot C3 \cdot C4}$$

Symbol	Description	Value	Units
C3	Loop Filter Capacitor	0.0740	nF
C4	Loop Filter Capacitor	0.0499	nFs
R2	Loop Filter Resistor	3.7040	kΩ
R4	Loop Filter Resistor	29.8734	kΩ

Chapter 23   Fundamentals of PLL Active Loop Filter Design

Introduction

The following several chapters have discussed passive loop filter designs. Passive loop filters are generally recommended over active filters for reasons of cost, simplicity, and in-band phase noise. The added in-band phase noise comes from the active device that is used in the loop filter. However in cases where the VCO requires a higher tuning voltage than the PLL charge pump can operate, active filters are necessary. VCOs with high voltage tuning requirements are most common in broadband tuning applications, such as those encountered in cable TV tuners. It is also commonly required for low noise or high power VCOs.

With older styles of phase detectors, before the charge pump PLL, active filters were used in order to obtain a zero steady-state phase error and infinite pull-in range. However, this is not a good reason to use an active filter with a charge pump PLL, since the charge pump PLL always attains these characteristics with a passive filter.

Many of the concepts presented in this chapter are analogous to those in passive loop filter design. The solution for the time constants is identical, however the solution of components from those time constants is not the same, since the active device does provide isolation for the higher stages. The concepts for loop bandwidth, phase margin and pole ratios all apply. It is generally recommended to use at least a third order filter, since the added pole reduces the phase noise of the active device.

Types of Active Filters

The two basic classes of active filters are those using the differential phase detector outputs and those that use the charge pump output pin. For each of these two basic classes, there are also different variations for the loop filter topology. Since most of the concepts in this chapter are not applicable to the approach involving the differential phase detector outputs, this case is treated in a separate chapter.

The other approaches presented all involve using active devices to boost the charge pump output voltage. One such way involves simply adding a gain stage before the VCO. Other approaches involve putting components in the feedback path of the active loop filter device.

Regardless of the approach used, there is usually a phase inversion introduced, which can be negated by reversing the polarity of the charge pump. There is also isolation added, which allows a larger capacitor to be chosen next to the VCO to reduce the impact of the VCO input capacitance and loop filter resistor noise.

The Pole Switching Trick

With passive filters, *T31<1* is a constraint for real component values. However, with active filters, *T3=T1* is a perfectly valid condition. For optimal spurious attenuation, this condition should be applied. However, often the op-amp noise is a larger consideration than the spurs. Because the pole *T3* comes after the op-amp, it offers more filtering of the op-amp noise.

From this perspective, it makes sense to make *T31 > 1*. In the case of a passive filter, changing the poles *T3* and *T1* have no impact on the loop filter components. In the case of an active filter, switching these two poles does not impact the loop parameters, but does change the loop filter components and also impacts how the op-amp noise is filtered. Making *T31>1* is effectively switching these poles.

Simple Gain Approach

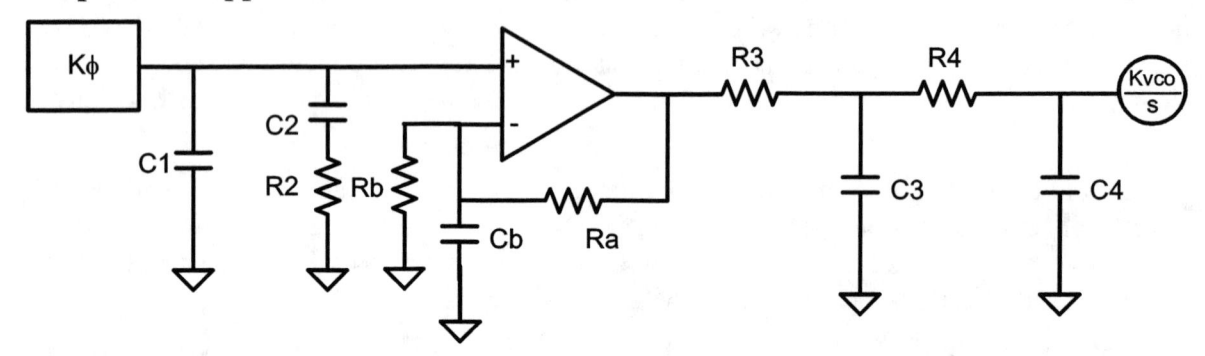

Figure 23.1 *An Active Filter Using the Simple Gain Approach*

This approach involves placing an op-amp in front of the VCO. The advantage of this approach is that it is very intuitive and commonly used. The disadvantage is that it does not optimally center the charge pump and it multiplies the op-amp noise. Since the op-amp generates noise, it is generally recommended to use a third or higher order filter to reduce the op-amp noise, even if the spurs do not benefit much from it.

The gain of *A* is produced by using an op-amp in an non-inverting configuration. The resistor **Rx** is selected to be large enough so that the current consumption is not excessive. However, choosing **Rx** excessively large could lead to problems due to the resistor thermal noise. If thermal noise is a concern, the capacitor **Cb** can be use to greatly reduce it. The gain, *A*, is always negative and is calculated as:

$$A = -(1 + Ra/Rb) \tag{23.1}$$

Feedback Approaches

The problem with the simple gain approach is that the op-amp noise is multiplied by the gain of the gain stage. Feedback approaches put part of the loop filter components in the feedback path of the op-amp and eliminate this problem. At the non-inverting input of the op-amp, it is necessary to establish a fixed voltage, called the bias voltage. Because this bias voltage is fixed, the charge pump output voltage can be held at a fixed voltage, which is usually half of the charge pump supply voltage. Because this voltage is fixed, spurs should be lower. The only disadvantage of the feedback approaches is that some of them require that the op-amp slew rate is sufficiently fast, but this can be avoided by using the slow slew rate approach.

The bias voltage is established by using a simple resistive divider. If the resistors are chosen too small, there will be excessive current consumption. If they are chosen too large, then there will be excessive resistor noise. In most cases, the resistors will be equal to bias the charge pump output at half of the supply. One trick that can be used is to use a shunt capacitor, **Cb**, in order to reduce this noise. The bias voltage is calculated from the bias resistors as follows:

$$Vbias = \frac{Vp \cdot Rb}{Ra + Rb} \qquad (23.2)$$

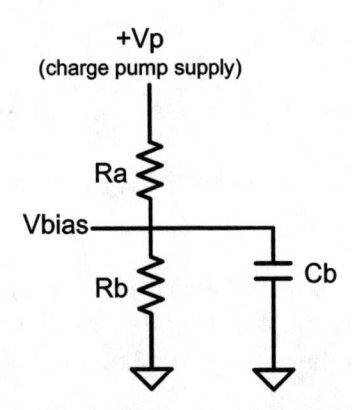

Figure 23.2 *How to Establish a Bias Voltage*

Standard Feedback Approach

This approach involves putting the components **C1**, **C2**, and **R2** in the feedback path of an op-amp. Additional filtering stages are added after the op-amp. This approach is generally superior to the simple gain approach because it allows the charge pump voltage to be centered at half the charge pump supply, for lower and more predictable spur levels.

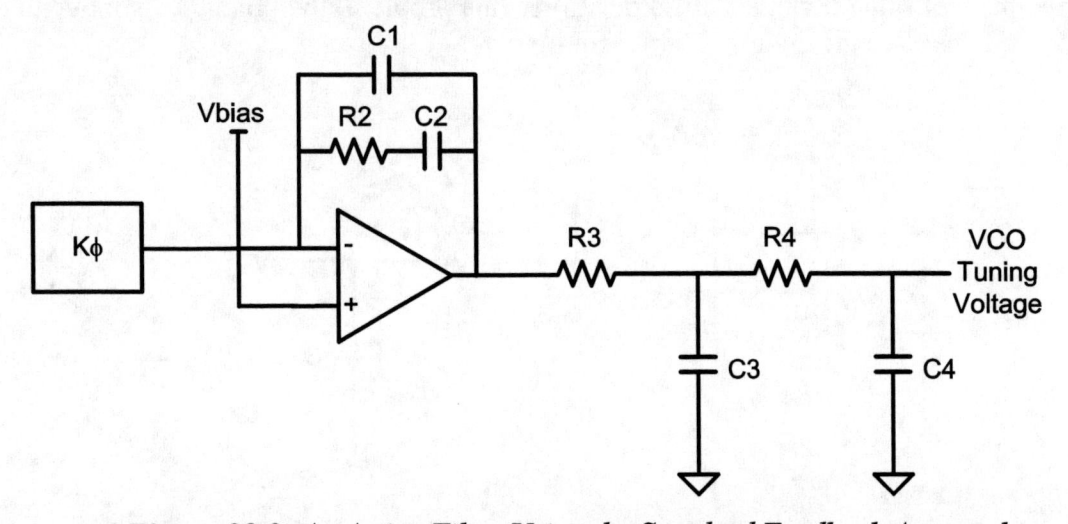

Figure 23.3 *An Active Filter Using the Standard Feedback Approach*

Alternative Feedback Approach

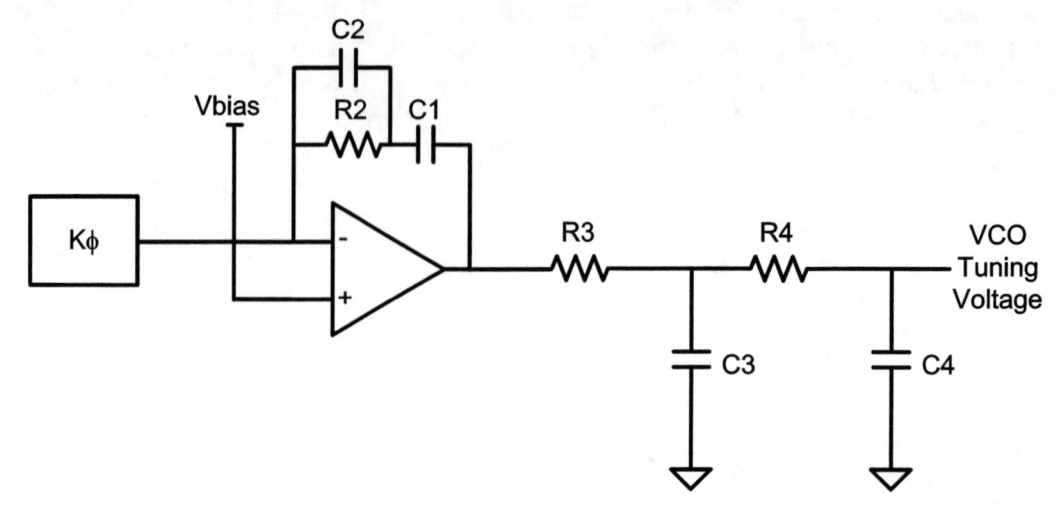

Figure 23.4 *An Active Filter Using the Alternative Feedback Approach*

This approach is very similar to the standard feedback approach, except that the topology is slightly changed. The only possible advantages or disadvantages of this approach would be a consequence of the fact that the actual calculated component values will be different.

Slow Slew Rate Approach

One of the problems with the standard feedback approach is that the charge pump output is presented directly to the op-amp. This puts requirements on the slew rate of the op-amp because the correction pulses from the charge pump are very fast. If the op-amp is not fast enough, then an AC waveform will be generated on the tuning line. Depending on how high the comparison frequency is, the additional filtering after the op-amp might be able to handle this, but often it cannot. In order to fix this problem, the pole, *T1*, is moved before the op-amp to relive the op-amp of this requirement.

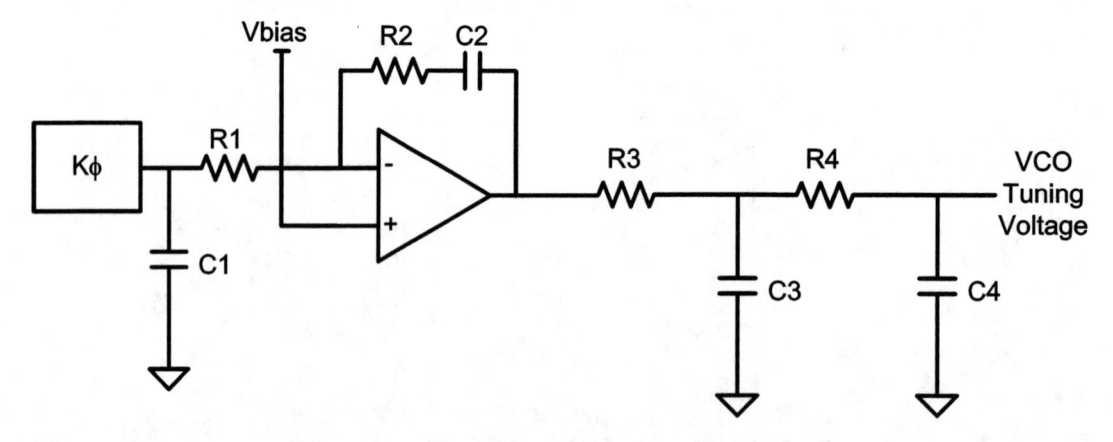

Figure 23.5 *Slow Slew Rate Modification to Standard Feedback Approach*

Using Transistors for the Standard and Alternative Feedback Approaches

For either of the feedback approaches, transistors can be used to replace the op-amp in order to reduce the cost and the noise. For the approach presented here, the transistors can only sink current, so a pull-up resistor, **Rpp**, is required. There is a gain introduced by the transistors, and this gain is dependent on **Vpp**, **Rpp**, and the tuning voltage. There are no formulas presented for this gain, so tinkering is required. Increasing **Vpp** or **Rpp** increases the gain, but the relationship is nonlinear. Decreasing the tuning voltage also increases the gain. Because this gain is hard to calculate, it is recommended to use a PLL with many charge pump settings to compensate for this variation in the gain. If there is too much gain, the circuit will become unstable. The choice of **Rpp** is design and possibly transistor specific, but **Rpp = 5.6 KΩ** is a good starting value. A popular choice for **Vpp** is 30 volts. Choosing this resistor too large will cause the circuit to be unstable and the carrier to dance around the frequency spectrum. Choosing it too small will cause excessive current consumption since **Vpp** is grounded through the resistor **Rpp** when the transistors turn on. This particular design has been built and tested to 30 volt operation. The 220 Ω resistor sets the bias voltage for the charge pump output. The optional 20 KΩ resistor may reduce the phase noise. The KΩ resistor limits the sink current. All of these values can be tweaked in order to improve performance. The main challenge in this design is knowing what gain the transistors give.

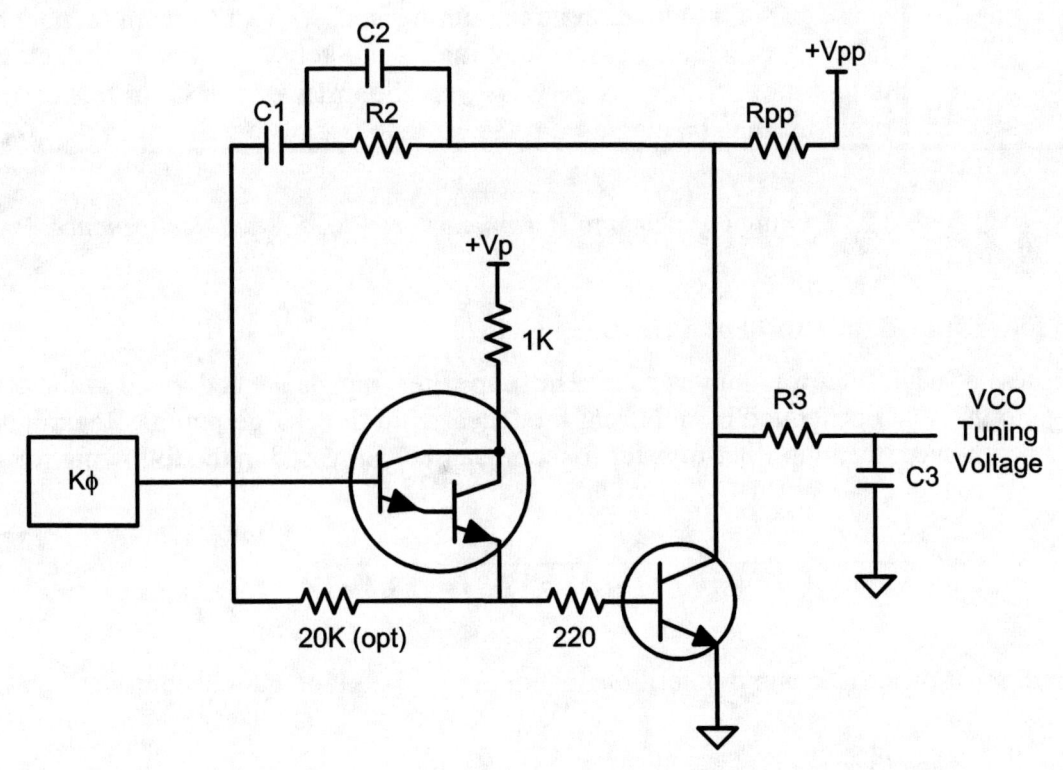

Figure 23.6 *Third Order Alternative Feedback Active Filter Using Transistors*

Choosing the Right Op-Amp

The choice of the correct op-amp is somewhat of an art. The table below summarizes how various parameters impact the system performance.

Parameter	Impact on PLL System
Offset Voltage	This has no impact on system performance.
Noise Voltage	This is very important and has a large impact on phase noise, especially at and outside the loop bandwidth.
Noise Current	This is also very important and has a large impact on phase noise.
Input Rails	In order to avoid a negative supply, it is preferable that the negative supply rail is small. For instance, if the PLL supply is 3 volts, and the negative supply rail is 4 volts, this forces the use of a negative supply. The positive supply rail is much less important.
Output Rails	The op-amp output voltage needs to be able to tune the VCO. The biggest thing to watch for is the negative output rail. If this is too large, it could force one to use a negative supply.
Slew Rate	This can impact both spurs and lock time. For spurs, if the standard or alternative feedback approach is used, if the slew rate is too slow, it can cause an AC modulation on the tuning line, which results in higher spurs. In terms of lock time, the slew rate is unlikely to degrade the lock time, unless the lock time is very fast, on the order of 5 μS. In this case, the peak time can be increased if the op-amp is too slow.

Table 23.1 *Impact of Op-Amp Parameters on PLL System Performance*

Loop Filter Impedance and Forward Loop Gain

Regardless of what filter topology is used, The loop filter impedance is defined as the output voltage to the VCO generated by a current produced from the charge pump. Regardless of the approach used, the loop filter transfer function can be expressed in the following form:

$$Z(s) = \frac{1+s \cdot T2}{s \cdot A0} \cdot \frac{A}{(1+s \cdot T1) \cdot (1+s \cdot T3) \cdot (1+s \cdot T4)} \qquad (23.3)$$

Assuming that the charge pump polarity is inverted, the open loop gain becomes:

$$G(s)/N = -\frac{K\phi \cdot Kvco \cdot A}{\omega^2 \cdot N} \cdot \frac{1+s \cdot T2}{s \cdot A0 \cdot (1+s \cdot T1) \cdot (1+s \cdot T3) \cdot (1+s \cdot T4)} \qquad (23.4)$$

	Simple Gain Approach	Feedback Approaches		Slow Slew Rate Approach
		Standard Feedback	Alternative Feedback	
T1	$\dfrac{C1 \cdot C2 \cdot R2}{C1+C2}$	$\dfrac{C1 \cdot C2 \cdot R2}{C1+C2}$	$C2 \cdot R2$	$C1 \cdot R1$
T2	$C2 \cdot R2$	$C2 \cdot R2$	$R2 \cdot (C1+C2)$	$C2 \cdot R2$
T3	$\dfrac{(C3 \cdot R3 + C4 \cdot R3 + C4 \cdot R4) + \sqrt{(C3 \cdot R3 + C4 \cdot R3 + C4 \cdot R4)^2 - 4 \cdot C3 \cdot C4 \cdot R3 \cdot R4}}{2}$			
T4	$\dfrac{(C3 \cdot R3 + C4 \cdot R3 + C4 \cdot R4) - \sqrt{(C3 \cdot R3 + C4 \cdot R3 + C4 \cdot R4)^2 - 4 \cdot C3 \cdot C4 \cdot R3 \cdot R4}}{2}$			
A0	$C1+C2$	$C1+C2$	$C1$	$C2$
A	$1 + Ra/Rb$	-1		

Table 23.2 *Filter Parameters as they Relate to the Filter Components*

Calculating the Loop Filter Components

Solving for the time constants

The first step in calculating the loop filter components is calculating the time constants. This is done in exactly the same way that it was done in the case of a passive filter, and is therefore not shown again in this chapter. Once the time constants are known, the loop filter components can be calculated from these time constants.

Solving for A0

The first step in solving for the components is determining the value of *A0*. This can be found by setting the open loop gain equal to one at the loop bandwidth.

$$A0 = \frac{K\phi \cdot Kvco \cdot A}{\omega c^2 \cdot N} \cdot \sqrt{\frac{1+\omega c^2 \cdot T2}{(1+\omega c^2 \cdot T1) \cdot (1+\omega c^2 \cdot T3) \cdot (1+\omega c^2 \cdot T4)}} \qquad (23.5)$$

Solving for the Components

Once that *A0* is found, the other components can be found using the Table 23.4. For a third order loop filter, *C3* should be at least four times the VCO input capacitance and at least *C1/5*. For a fourth order loop filter, *C4* should be at least this stated limit above.

	Simple Gain Approach	Feedback Approaches		Slow Slew Rate
		Standard	Alternative	
$C1$	$A0 \cdot \dfrac{T1}{T2}$	$A0 \cdot \dfrac{T1}{T2}$	$A0$	*Free to choose. Suggest 1000 pF.*
$C2$	$A0 \cdot \left(1 - \dfrac{T1}{T2}\right)$	$A0 \cdot \left(1 - \dfrac{T1}{T2}\right)$	$A0 \cdot \dfrac{T1}{T2 - T1}$	$A0$
$R2$	$\dfrac{T2}{C2}$	$\dfrac{T2}{C2}$	$\dfrac{T2}{C1 + C2}$	$\dfrac{T2}{C2}$
Third Order Filter Components				
$C3$	*Choose C3 at least 4X the VCO input capacitance and at least 200 pF.*			
$R3$	$\dfrac{T3}{C3}$			
Fourth Order Components				
$C4$	*Choose C4 at least 4X the VCO input capacitance and preferably at least 220 pF. Also make sure that this yields realistic values for C3.*			
$C3$	$C4 \cdot \dfrac{4 \cdot T3 \cdot T4}{(T3 - T4)^2}$			
$R3$	$\dfrac{T3 + T4}{2 \cdot (C3 + C4)}$			
$R4$	$\dfrac{T3 + T4}{2 \cdot C4}$			

Table 23.3 *Loop Filter Component Values Computed from Time Constants*

Measured Phase Noise Results

Phase noise is a critical consideration in designing active filters. In order to measure this, several loop filters were designed using the same PLL and VCO. A passive filter of the same loop bandwidth was designed and the noise from this was subtracted. What these results show is that it really is worthwhile to put additional filtering after the op-amp and the phase noise tends to be most degraded near the loop bandwidth. The loop bandwidth used was 10 kHz in all cases and the noise voltage of the op-amp used was 6 nV/sqrt(Hz).

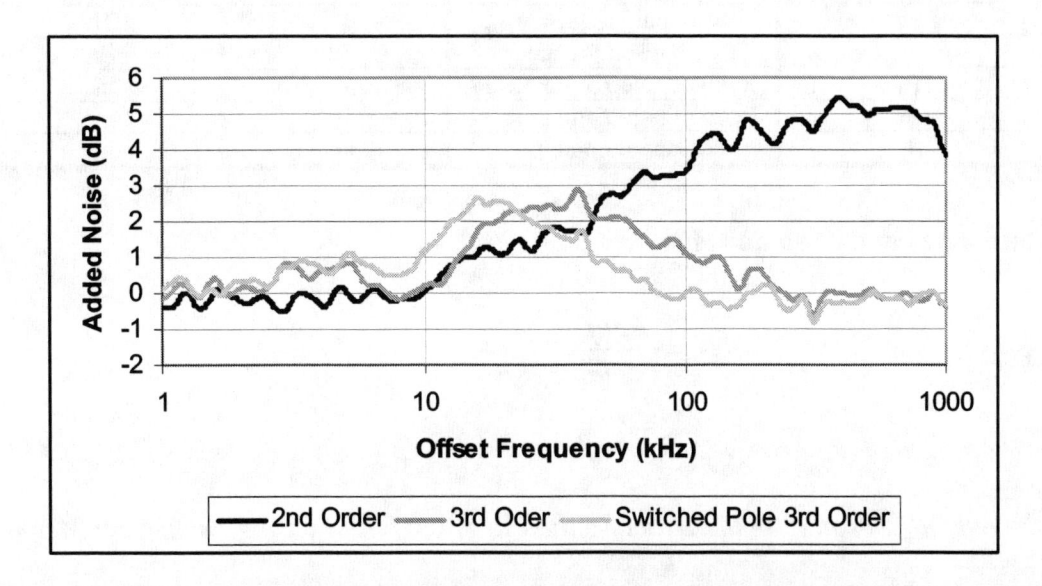

Figure 23.7 *Added Phase Noise Due to an Op-Amp*

Conclusion

The equations for active loop filter design have been presented. Active filters are necessary when the charge pump can not operate at high enough voltages to tune the VCO and can also help reduce the ill effects of the VCO input capacitance. The choice of the op-amp is somewhat of an art. One has to balance the input and output rails, bias currents, noise voltage, and noise current. For instance, the LMH6624 has excellent noise performance, but very high input bias currents. If a fractional PLL is used, it might be possible to make the comparison frequency high enough to tolerate these higher bias currents. If an integer PLL is used, then one needs to choose an op-amp with lower bias currents. The LMV751 has good noise and bias currents, but only goes to 5.5 volts. The OP27 has good noise and bias currents, but the input rails requires the use of a negative supply. A poor choice for the op-amp could easily increase the phase noise by 10 dB, while a good choice would probably increase the phase noise by a couple dB.

References

I had useful conversations with John Bittner and Eric Eppley regarding active filter design.

Appendix A: An Active Filter Design Example

Symbol	Description	Value	Units
Fc	Loop Bandwidth	20	kHz
ϕ	Phase Margin	47.8	degrees
γ	Gamma Optimization Parameter	1.115	none
$K\phi$	Charge Pump Gain	5	mA
Kvco	VCO Gain	44	MHz/V
Fout	Output Frequency	2441	MHz
Fcomp	Comparison Frequency	500	kHz
T31	Ratio of pole T3 to Pole T1	1/0.4 = 2.5	none
T41	Ratio of pole T4 to Pole T1	0.4	none
A	Gain of op-amp	1 and 3	none

Calculate Poles and Zero and $A0$

$$N = \frac{Fout}{Fcomp} \tag{23.6}$$

$$\omega c = 2 \bullet \pi \bullet Fc \tag{23.7}$$

T1 is the only unknown. Use the Exact Method to Solve for *T1* Using Numerical Methods

$$\phi = tan^{-1}\left(\frac{\gamma}{\omega c \bullet T1 \bullet (1+T31)}\right) - tan^{-1}(\omega c \bullet T1) \tag{23.8}$$
$$- tan^{-1}(\omega c \bullet T1 \bullet T31) - tan^{-1}(\omega c \bullet T1 \bullet T31 \bullet T43)$$

$$T3 = T1 \bullet T31 \tag{23.9}$$

$$T4 = T1 \bullet T43 \tag{23.10}$$

$$T2 = \frac{\gamma}{\omega c^2 \bullet (T1+T3+T4)} \tag{23.11}$$

$$A0 = \frac{K\phi \bullet Kvco \bullet A}{\omega c^2 \bullet N} \bullet \sqrt{\frac{1+\omega c^2 \bullet T2}{(1+\omega c^2 \bullet T1)\bullet(1+\omega c^2 \bullet T3)\bullet(1+\omega c^2 \bullet T4)}} \tag{23.12}$$

Symbol	Description			Value	Units
N	N Counter Value			4882	none
ωc	Loop Bandwidth			1.2566×10^5	rad/s
$T1$	Loop Filter Pole			7.0269×10^{-7}	s
$T2$	Loop Filter Zero			2.2330×10^{-5}	s
$T3$	Loop Filter Pole			1.7567×10^{-6}	s
$T4$	Loop Filter Pole			7.0269×10^{-7}	s
$A0$	Lowest Order Loop Filter Coefficient for All Feed Back Approaches			8.2367	nF
	Lowest Order Loop Filter Coefficient for Simple Approach with Gain of 3			24.7100	nF
	Simple Gain Approach	Feedback Approaches		Slow Slew Rate	
		Standard	Alternative		
A	3	1			
$C1$	0.8593 nF	0.2984 nF	8.1604 nF	1.0000 nF	
$C2$	23.5859 nF	7.8620 nF	0.3098 nF	8.1604 nF	
$R2$	0.9433 kΩ	2.8300 kΩ	2.6268 kΩ	2.8300 kΩ	
$C4$	0.5600 nF				
$C3$	0.5079 nF				
$R3$	1.1048 kΩ				
$R4$	2.1069 kΩ				

Chapter 24 Active Loop Filter Using the Differential Phase Detector Outputs

Introduction

This chapter investigates the design and performance of a loop filter designed using the differential phase detector outputs, ϕr and ϕn. In general, modern PLLs have excellent charge pumps on them and it is generally recommended not to bypass it. In doing so, all models concerning phase noise and spurs presented in this book become invalid. In fact, most modern PLLs do not have these differential phase detector outputs. For those who insist on bypassing the charge pump and using these differential outputs, this chapter included.

Loop Filter Topology

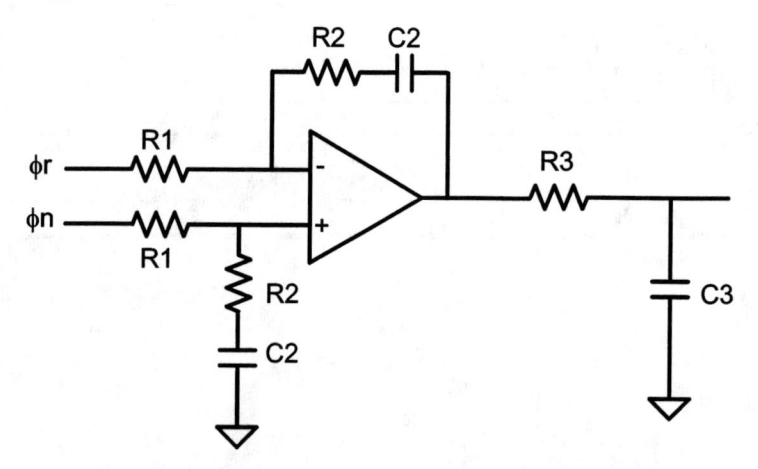

Figure 24.1 *Active Filter Topology Used*

The transfer function of the filter is given by:

$$Z(s) = \frac{1 + s \bullet T2}{s \bullet T \bullet (1 + s \bullet T1)} \qquad (24.1)$$

where

$$T2 = R2 \bullet C2 \qquad (24.2)$$

$$T1 = R3 \bullet C3 \qquad (24.3)$$

$$T = R1 \bullet C2 \qquad (24.4)$$

The open loop response is given by:

$$\frac{G(s)}{N} = \frac{Kv \cdot Kvco \cdot (1+s \cdot T2)}{N \cdot T \cdot s^2 \cdot (1+s \cdot T1)} \qquad (24.5)$$

From the chapter on a second order passive filter, this transfer function has many similarities. If the following substitutions are applied to expression for the open loop response for the second order filter, then the result is the transfer function for this loop filter topology. In these equations, **Kv** represents the maximum voltage output level of the phase detector outputs.

$$T \Rightarrow A0 \qquad (24.6)$$
$$Kv \Rightarrow K\phi$$

The case where **R3 = C3 = 0** presents a special case and has different equations, but is a topology that is sometimes used. This approach will be referred to as the alternative approach, and the case where **T1>0** will be referred to as the standard approach. In either case, the equations for the time constants and filter components are shown in Table 24.1 .

Component	Standard Approach	Alternative Approach
T1	$T1 = \dfrac{\sec(\phi) - \tan(\phi)}{\omega c}$	0
T2	$T2 = \dfrac{1}{\omega c^2 \cdot T1}$	$\omega c \cdot \tan\phi$
T	$T = \dfrac{Kv \cdot Kvco}{N \cdot \omega c^2} \cdot \sqrt{\dfrac{1+\omega c^2 \cdot T2^2}{1+\omega c^2 \cdot T1^2}}$	$T = \dfrac{Kv \cdot Kvco}{N \cdot \omega c^2 \cdot \cos\phi}$
C2	Choose this value	Choose this value
R2	$\dfrac{T2}{C2}$	$\dfrac{T2}{C2}$
R1	$\dfrac{T}{C2}$	$\dfrac{T}{C2}$
C3	Choose this at least four times the VCO input capacitance. Preferably at least 220 pF.	0
R3	$\dfrac{T3}{C3}$	0

Table 24.1 *Loop Filter Time Constants and Component Values*

Conclusion

This chapter has presented design equations that can be used with the differential phase detector outputs. This approach is generally not recommended, because it requires an op-amp and most PLLs do not have these differential output pins. The reader should also be very aware of the states of the outputs. For instance, when this type of loop filter is used with National Semiconductor's LMX2301/05/15/20/25 PLLs, it is necessary to invert either *ϕr* or *ϕn*.

There are other approaches to loop filter design using these differential outputs. One such approach is to omit the components **R3** and **C3**. In this case, **T1** becomes zero and **T2** becomes *ωc•tan(ϕ)*. This topology is more popular with older PLL designs than newer ones.

The lock time can be predicted with a formula, but the phase noise and spurs for this filter differ than those in a passive filter. The ***BasePulseSpur*** and ***1HzNoiseFloor*** are different, since the charge pump has been bypassed.

Reference

AN535   *Phase-Locked Loop Design Fundamentals*   Motorola Semiconductor Products, 1970

Chapter 25 Impact of Loop Filter Parameters and Filter Order on Spur Levels

Introduction

It has been shown that the reference spur levels are directly related to the spur gain, whether they are leakage or pulse dominated. This chapter investigates methods of minimizing the spur gain under various conditions. First, it will be shown why choosing all the pole ratios (**T31**, **T41**,...) equal to one always yields the lowest spur gain filter. Then, the impact of other loop filter design parameters on the spur gain will also be investigated. Recall that in a previous chapter, the impact of various parameters was analyzed in the case that the loop filter was not redesigned. In this chapter, it will be assumed that the loop filter is redesigned. For instance, having a bigger VCO gain increases spur levels if the loop filter is not redesigned. But, it turns out that it has no impact if the loop filter is redesigned to have the same loop bandwidth.

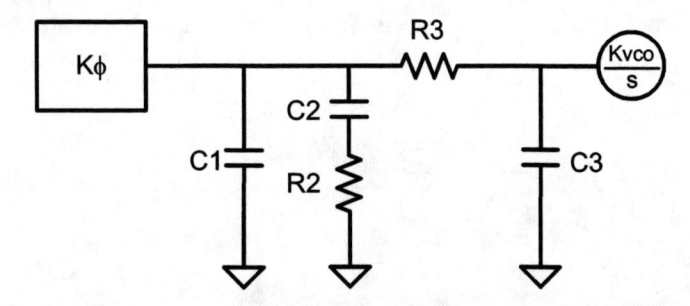

Figure 25.1 *Basic Passive Loop Filter Topology*

Minimization of Spur Gain

Since the spur levels relate directly to the spur gain of the PLL, the problem is therefore reduced to minimizing the spur gain under the constraints of a constant loop bandwidth and phase margin. The poles of the filter will be represented by Ti, $(i = 1, 3, 4, \ldots k)$. Note that $T2$ is the zero of the filter and therefore the index skips over two. The filter order is k, which is assumed to be greater than two. $T1(i)$ is intended to mean the ratio of pole Ti to the pole $T1$. This number can range from zero to one. Note that $T1(1) = 1$. The spur gain at any frequency can be expressed as:

$$|G(\omega)| = \frac{K\phi \cdot Kvco}{A0 \cdot \omega^2} \cdot \sqrt{\frac{1 + \omega^2 \cdot T2^2}{\prod_{i=1,3,4,\ldots k}(1 + \omega^2 \cdot Ti^2)}} \qquad (25.1)$$

However $A0$ is not constant. Recall:

$$A0 = \frac{K\phi \cdot Kvco}{N \cdot \omega c^2} \cdot \sqrt{\frac{1+\omega c^2 \cdot T2^2}{\prod_{i=1,3,4,...k}(1+\omega c^2 \cdot Ti^2)}} \qquad (25.2)$$

Substituting this in gives the following expression for $G(s)$:

$$|G(s)| = N \cdot \frac{\omega c^2}{\omega^2} \cdot \sqrt{\frac{1+\omega^2 \cdot T2^2}{1+\omega c^2 \cdot T2^2} \cdot \prod_{i=1,3,4,...k} \frac{(1+\omega c^2 \cdot Ti^2)}{(1+\omega^2 \cdot Ti^2)}} \qquad (25.3)$$

The above equation eliminates all of the component values from the equations, but still leaves the time constants to be calculated. However, there are three approximate equations that relate the time constants to known design parameters. It therefore follows that the spur gain can be expressed uniquely in terms of design parameters.

$$T1 = \frac{sec\phi - tan\phi}{\omega c \cdot \sum_{i=1,3,4,...k} T1(i)} \qquad (25.4)$$

$$Ti = T1 \cdot T1(i) = \frac{sec\phi - tan\phi}{\omega c \cdot \sum_{i=1,3,4,...k} T1(i)} \cdot T1(i) \qquad (25.5)$$

$$T2 = \frac{1}{\omega c^2 \cdot \sum_{i=1,3,4,...k} Ti} = \frac{1}{\omega c \cdot (sec\phi - tan\phi)} \qquad (25.6)$$

Substituting (25.4), (25.5), and (25.6) into (25.3) yields the spur gain in terms of design parameters.

$$|G(s)| = \frac{N}{r^2} \cdot \sqrt{\frac{r^2+x^2}{1+x^2} \cdot \prod_{i=1,3,4,...k} \left[\frac{\left(\sum_{j=1,3,4,...k} T1(j)\right)^2 + T1(i)^2 \cdot x^2}{\left(\sum_{j=1,3,4,...k} T1(j)\right)^2 + T1(i)^2 \cdot x^2 \cdot r^2} \right]} \qquad (25.7)$$

The following terms are defined above:

$$x = sec\phi - tan\phi \qquad (25.8)$$

$$r = \frac{Spur\ Frequency}{Loop\ Bandwidth} = \frac{Fspur}{Fc} \qquad (25.9)$$

Since there is a leading $\frac{1}{r^2}$ term, it should be clear that the spur gain is minimized for the smallest values of r, which corresponds to minimizing the loop bandwidth. Some other things that are a little less obvious are the relationship of spur gain to the parameter x and the relationship of spur gain to the poles ratios of the filter. Since r can be assumed to be greater than one, it can be shown that (25.7) is a decreasing function in $T1(i)$ for $i=1,3,...k$. However, these pole ratios cannot exceed one, since $T1$ is by definition the largest pole. From this observation comes the fundamental result that for minimum spur levels, the pole ratios should all be chosen to be one. However, choosing all of the pole ratios to be one can yield a loop filter with a very small capacitor next to the VCO, which can be impacted by the VCO input capacitance. In the case of using the improved design equations for a fourth order filter, this capacitor would be zero. So there is often a good reason why the pole ratios should be chosen less than one.

One can reason from (25.7) that this function is a decreasing function of $|x|$, because if $r > 1$, this makes each one of the fractional parts decreasing functions in $|x|$, therefore the whole function is decreasing in $|x|$. So, for the minimum spur levels, this is equivalent to minimizing (25.8). Going through this exercise shows that this function is an increasing function in ϕ in the interval from 0 to 90 degrees, and therefore minimizing the spur gain corresponds to minimizing the phase margin. However, in practice, the impact of changing the phase margin typically does not have much of an impact on spurs. In the chapter on lock time, the second order function implies that lower phase margins also yield faster lock times. However, computer simulations using the 4th order model show that the phase margin that yields the fastest lock time is usually about 48 degrees. Therefore, it makes sense to design for a phase margin near 48 degrees, because this gives more freedom to adjust the loop bandwidth, which has a far greater impact on spur levels than phase margin.

Symbol	Description	Leakage Dominated Spurs	Mismatch Dominated Spurs
CP_{tri}	Charge Pump Leakage,	$20 \cdot \log(CP_{tri})$	N/A
CP_{mm}	Charge Pump Mismatch	N/A	Correlated to $\|CP_{mm} - Constant\|$
N	N Counter Value	$20 \cdot \log(N)$	$20 \cdot \log(N)$
$Kvco$	VCO Gain	Independent	Independent
Fc	Loop Bandwidth	$40 \cdot \log(Fc)$	$40 \cdot \log(Fc)$
$Fcomp$	Comparison Frequency	$-40 \cdot \log(Fcomp)$	$-40 \cdot \log(Fcomp)$
r	$=Fcomp/Fc$	$-40 \cdot \log(r)$	$-40 \cdot \log(r)$
$K\phi$	Charge Pump Gain	$-10 \cdot \log(K\phi)$	Independent
ϕ	Phase Margin	Weak Inverse Correlation	
$T31$	Ratio of $T3$ to $T1$	Inverse Correlation	

Table 25.1 *Reference Spur Gain vs. Various Loop Filter Parameters*

From Table 25.1, it follows that the loop bandwidth, comparison frequency, and N value have the largest influence on the spur level. If one considers the ratio of the comparison frequency to the loop bandwidth, then this is a rough indicator. The N value is also relevant, but is related to the comparison frequency. Larger charge pump gains yield lower leakage dominated spurs, because they yield larger capacitor values in the loop filter. The reader should be very careful to realize that these values assume that the loop filter is redesigned and optimized. If the loop filter is not redesigned, then the results will be very different. These results were derived in a previous chapter.

(25.7), (25.8), and (25.9) show that the spur gain of a third order filter is approximated by:

$$SG = 20 \cdot \log(N) - 40 \cdot \log(r) \qquad (25.10)$$
$$+ 10 \cdot \log \left| \frac{r^2 + x^2}{1 + x^2} \cdot \frac{(1+T31)^2 + T31^2 \cdot x^2}{(1+T31)^2 + T31^2 \cdot x^2 \cdot r^2} \cdot \frac{(1+T31)^2 + x^2}{(1+T31)^2 + x^2 \cdot r^2} \right|$$

So the **$20 \cdot \log(N)$** term shows the clear dependence on N, and therefore, Table 25.2 assumes an N value of one, to which this **$20 \cdot \log(N)$** must be added. Note that these equations assume that the filter is redesigned. If this is not the case, then it turns out that the spurs are not impacted much by the N value. The phase margin and r values are given. From this, go and find the main block, and then find the corresponding value of the **$N=1$** normalized spur gain. To this, add **$20 \cdot \log(N)$** to get the total spur gain.

		r										
		3	5	10	15	20	25	50	100	200	500	1000
T31 = 0	φ=30	-15.4	-23.6	-35.3	-42.3	-47.3	-51.2	-63.2	-75.2	-87.3	-103.2	-115.2
	φ=40	-14.1	-22.0	-33.6	-40.5	-45.5	-49.3	-61.3	-73.4	-85.4	-101.3	-113.4
	φ=50	-12.9	-20.3	-31.5	-38.4	-43.3	-47.2	-59.2	-71.2	-83.3	-99.2	-111.2
	φ=60	-11.7	-18.4	-29.1	-35.9	-40.8	-44.6	-56.5	-68.6	-80.6	-96.5	-108.6
	φ=70	-10.6	-16.5	-26.1	-32.5	-37.3	-41.1	-52.9	-64.9	-77.0	-92.9	-104.9
T31 = .25	φ=30	-14.9	-23.5	-37.5	-46.8	-53.7	-59.3	-77.0	-94.9	-113.0	-136.8	-154.9
	φ=40	-13.5	-21.6	-34.7	-43.6	-50.3	-55.7	-73.2	-91.1	-109.2	-133.0	-151.1
	φ=50	-12.3	-19.6	-31.8	-40.1	-46.6	-51.8	-69.0	-86.8	-104.8	-128.6	-146.7
	φ=60	-11.2	-17.7	-28.7	-36.3	-42.3	-47.2	-63.8	-81.5	-99.4	-123.2	-141.3
	φ=70	-10.3	-15.9	-25.3	-32.0	-37.3	-41.8	-57.2	-74.3	-92.1	-115.9	-134.0
T31 = .50	φ=30	-14.8	-24.0	-39.2	-49.1	-56.3	-62.0	-79.9	-97.9	-116.0	-139.8	-157.9
	φ=40	-13.4	-21.7	-36.0	-45.5	-52.6	-58.3	-76.1	-94.1	-112.1	-136.0	-154.0
	φ=50	-12.1	-19.5	-32.5	-41.6	-48.5	-54.0	-71.7	-89.6	-107.7	-131.5	-149.6
	φ=60	-11.0	-17.4	-28.9	-37.2	-43.8	-49.1	-66.4	-84.3	-102.3	-126.1	-144.2
	φ=70	-10.2	-15.7	-25.1	-32.2	-38.0	-42.9	-59.4	-77.0	-94.9	-118.8	-136.8
T31 = .75	φ=30	-14.8	-24.2	-39.8	-49.8	-57.1	-62.8	-80.8	-98.8	-116.8	-140.7	-158.8
	φ=40	-13.3	-21.8	-36.4	-46.2	-53.4	-59.1	-76.9	-94.9	-113.0	-136.9	-154.9
	φ=50	-12.0	-19.5	-32.9	-42.2	-49.2	-54.8	-72.5	-90.5	-108.5	-132.4	-150.5
	φ=60	-11.0	-17.4	-29.0	-37.6	-44.3	-49.7	-67.2	-85.1	-103.1	-127.0	-145.0
	φ=70	-10.2	-15.6	-25.1	-32.3	-38.3	-43.3	-60.1	-77.8	-95.8	-119.6	-137.7
T31 = 1.0	φ=30	-14.8	-24.3	-39.9	-50.0	-57.3	-63.0	-80.9	-99.0	-117.0	-140.9	-159.0
	φ=40	-13.3	-21.8	-36.6	-46.3	-53.6	-59.2	-77.1	-95.1	-113.2	-137.0	-155.1
	φ=50	-12.0	-19.5	-32.9	-42.3	-49.4	-54.9	-72.7	-90.7	-108.7	-132.6	-150.7
	φ=60	-11.0	-17.3	-29.1	-37.7	-44.4	-49.8	-67.4	-85.3	-103.3	-127.2	-145.2
	φ=70	-10.2	-15.6	-25.1	-32.4	-38.4	-43.4	-60.3	-78.0	-96.0	-119.8	-137.9

Table 25.2 *Relative N=1 Normalized Spur Gains for a Third Order Filter*

Choosing the Right Filter Order

If one assumes 50 degrees phase margin and takes (25.7) and assumes that all the poles are equal, then the relative attenuation of a filter over a second order filter can be calculated. Some areas are darkly shaded to indicate that the loop filter order is too high and not practical.

		$r = Fspur/Fc$					
		1000	100	50	20	10	5
Loop Filter Order	3	39.4	19.5	13.5	6.0	1.4	-1.0
	4	74.2	34.3	22.5	8.2	0.6	-1.9
	5	105.9	46.1	28.6	8.5	-0.7	-2.7

Table 25.3 *Spur Improvement for Various Order Filters Above a Second Order Filter*

Although the table does contain some approximations, it does establish an upper estimate for the attenuation that can be achieved. Notice that when the comparison frequency is large relative to the loop bandwidth, there is much more advantage in building higher order filters. Of course in these cases, spurs are often not as much of an issue. The chart also implies that a third order loop filter (two poles) only makes sense if the comparison frequency is at least ten times the loop bandwidth. Although the maximum attenuation is for the case when $T1 = T3 = ... = Tk$, it sometimes makes sense to design for $T1 > T3 > ... > Tk$, in order to keep the capacitors large enough as to not be distorted by the VCO input capacitance and to better justify the approximations made.

Choosing $T31$ for a Third Order Filter

Although a larger $T31$ value always yields theoretically lower spur levels, there is a point at which increasing $T31$ yields diminishing returns for spur improvement, but the capacitor $C3$ approaches zero and the resistor $R3$ approaches infinity. Below is a table that shows what value of $T31$ is necessary to be 0.5 dB less than the theoretical maximum. This table was compiled for a gamma value of one and a phase margin of 50 degrees. The consequence of these findings is that they show that if one chooses $T31$ to be 62.2%, then most of the theoretical benefit of using a third order filter will be realized.

r=10	r=20	r=50	r=100	r=infinite
46.0%	58.6%	61.6%	62.0%	62.2%

Table 25.4 *Minimum T31 Value Needed to Be Within 0.5 dB of the Maximum Benefit*

Choosing $T31$ and $T43$ for a 4th Order Filter

For the sake of simplicity, $T43$ is defined as follows:

$$T43 = \frac{T41}{T31} \tag{25.11}$$

This requires a little bit more analysis than the third order filter. If one tries to get within 0.5 dB of the theoretical maximum spur benefit, then the component values will almost always be negative. It's not obvious what the trade-off is between $T31$, $T43$, spur gain, and the capacitor $C4$. Although a detailed theoretical analysis of this would be almost impossible, a computer simulation is much more feasible. The table below shows the range of $T31$ and $T43$ values that allow one to get within 2 dB of the theoretical maximum benefit of using a 4th order filter. The shaded areas indicate that there is no $T31$ value which can get the spur gain within 2 dB of the theoretical minimum value.

		r Value			
		r=10	r=20	r=50	r=100
T43 Value	0.1				
	0.2				
	0.3		71.3		
	0.4		59.2	95.9	
	0.5		52.0	71.5	75.0
	0.6		47.0	**60.5**	**62.7**
	0.7		43.2	53.9	55.4
	0.8		40.3	49.4	50.7
	0.9		38.0	46.1	47.2
	1.0		36.0	43.6	44.6

Figure 25.2 *T31 value as a percentage as a function of T43 and r values that attain a spur gain 2 dB less than the theoretical minimum.*

To better explain the above table, consider the following example with a 4th order loop filter and a comparison frequency of 200 kHz and a fixed loop bandwidth of 4 kHz. Assume a phase margin of 50 degrees and a gamma value of one. In this case, r=50 and a 4th order loop filter can theoretically improve spurs by 22.5 dB above a second order filter. If one designs for a *T31* value of 71.5% and a *T43* value of 50%, then one will achieve all but 2 db of this benefit. In other words, one can get a 20.5 dB benefit from using a 4th order loop filter over a 2nd order loop filter. However, one can achieve the same spur benefit choosing *T31* as 46.1% and *T43* as 90%. In fact, there is a whole continuous range of *T31* and *T43* values which fit this description. Which one is best? The optimal choice would be the one that maximizes the value for *C4*. By trial and error, the values in bold in the table have been found to maximize the value for *C4*. In other words, a *T31* value of 60.5% and a *T43* value of 60% would be the best choice for the table. In general, choosing *T31* = *T43* is not exactly the constraint that optimizes *C4*, but serves as a very good rule of thumb. Note that if *T31* and *T43* are chosen too large, then the component values will be negative. So an excellent starting point is to choose *T31* = *T43* = 50%. If the *r* value is 50 or higher, then perhaps this can be increased to 60%.

Comment Regarding Active Filters

For active filters, it actually makes sense to choose *T31*>100%. In this case, the results are the reciprocal of what has been obtained. For instance, the result that if one chooses *T31*=62.2% will sacrifice at most 0.5 dB in spurs relative to choosing *T31* = 100% corresponds to saying that if one chooses 100% < *T31* < 160.8% will sacrifice at most 0.5 dB in spurs.

Conclusion

This chapter investigated the impact of designing loop filters of higher than second order and when it makes sense to do so. One fundamental result is that the lowest reference spurs occur when the pole ratios are chosen equal to one. However, choosing all pole ratios equal to one can yield very small capacitor values next the VCO, which are easily be impacted by the VCO input capacitance. If one is designing a fourth order filter using the improved calculations, this would imply that *C4 = 0*. Another result was derived that illustrated that there is not much point to make *T31* much larger than 62.2% for a third order filter. In the case of a fourth order filter, this result is a little more complicated. When presented with a situation where the spur to be filtered is less than $1/10^{th}$ of the loop bandwidth, higher order filters do not help much, so it might make more sense to use the Fastlock feature or a switched mode filter.

Chapter 26 Optimal Choices for Phase Margin and Gamma Optimization Parameter

Introduction

This chapter investigates how to choose the phase margin and gamma optimization parameter, γ, in an optimal way. The gamma optimization parameter is the key to designing loop filters in an optimal way. It is possible to design two second order loop filters with the exact phase margin and loop bandwidth and still have one which has dramatically better lock time and spurs. The difference would be in the gamma optimization parameter. Many previous loop filter design techniques assume a gamma value of one, which is a good starting point, but there is further room for optimization. The optimal choice for gamma is dependent on the phase margin. For this reason, it is necessary to study the gamma optimization parameter and phase margin together in order to know the best values for these parameters to give the best spurs and lock time. For those applications where lock time is not critical, the optimal choice of phase margin to minimize peaking and RMS phase error is also discussed.

Definition of the Gamma Optimization Parameter

If one imposes the design constraint that the phase margin is maximized at the loop bandwidth, then this is equivalent to designing for a gamma value of one. Imposing this restriction yields the following equation:

$$\frac{1}{\omega c^2 \cdot T2} = \frac{1}{\omega c^2 \cdot T1} + \frac{1}{\omega c^2 \cdot T3} + \frac{1}{\omega c^2 \cdot T4} \qquad (26.1)$$

This can be approximated as:

$$\frac{1}{\omega c^2 \cdot T2} = \frac{1}{\omega c^2 \cdot (T1+T3+T4)} \qquad (26.2)$$

However, since choosing the phase margin to be optimized at the loop bandwidth is a good approximation to minimizing the lock time, but not the exact constraint, it makes sense to generalize this constraint. By introducing the variable, γ, but still keeping the equation in a similar form, one has a good idea of what values to try for this new variable. The new constraint can be stated as follows:

$$\frac{1}{\omega c^2 \cdot T2} = \frac{\gamma}{\omega c^2 \cdot (T1+T3+T4)} \qquad (26.3)$$

Eliminating and Normalizing Out Other Design Parameters

In the lock time chapter, recall that is was proven that, provided that only the loop bandwidth was changed, the lock time was inversely proportional to the loop bandwidth. What this means is that whatever choice of phase margin and gamma are optimal for one loop bandwidth, is also optimal for another loop bandwidth. The VCO gain, N value, and charge pump gain change the filter components, but have no impact on lock time, provided the loop filter is redesigned. So the only thing left to study is the pole ratios, phase margin, and gamma optimization factor. Now it will turn out that the pole ratios will have a small impact on the gamma parameter choice, and the phase margin will have the largest impact.

Results of Computer Simulations

It also turns out that the size of the frequency jump has a slight impact on the lock time, but this effect is minimal. So the approach is to assume fixed conditions for the frequency jump and tolerance, and then compile tables for the optimal gamma value based on computer simulations that cover all cases. Below are the conditions used to simulate the Gamma Parameters

Parameter	Value	Units
$K\phi$	5	mA
Kvco	20	MHz/Volt
Fc	10	kHz
Fcomp	200	kHz
Phi	Variable	Degrees
Frequency Jump	800 – 900	MHz
Frequency Tolerance for Lock Time	1	kHz
N	4500	n/a

Table 26.1 *Conditions for Simulations*

First Simulation: Impact of Gamma Value and Phase Margin on Lock Time

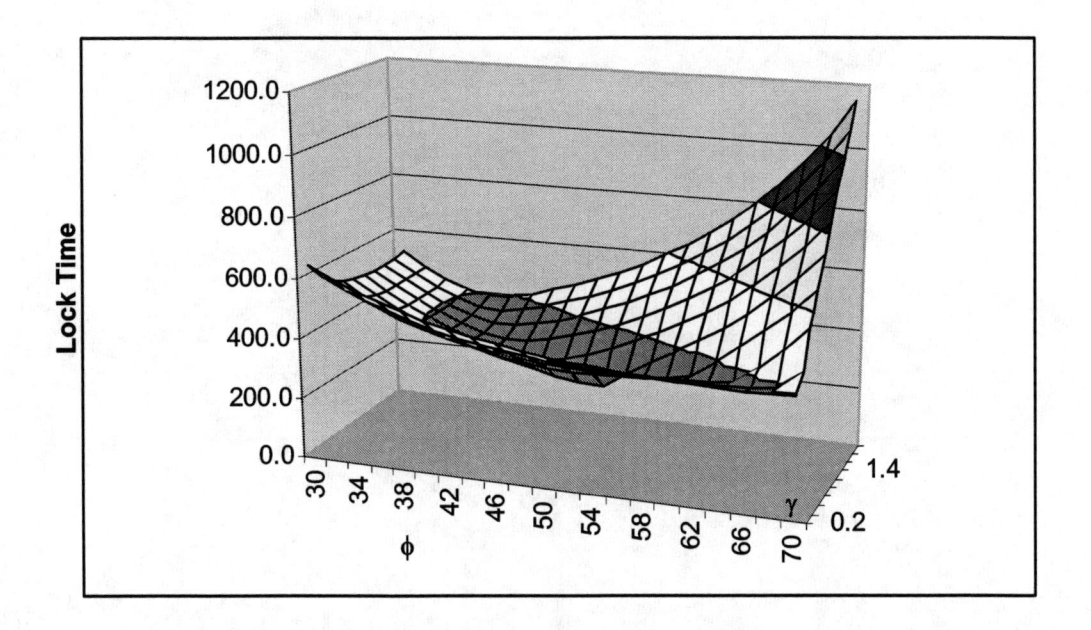

Figure 26.1   *Lock Time as a Function of Phase Margin and Gamma*

Figure 26.1 shows the lock time for this loop filter as a function of phase margin and the Gamma optimization parameter for a second order filter. There is a specific value of gamma and phase margin that minimize the lock time. Later in this chapter, this will be shown to be a phase margin of 50.8 degrees and a gamma value of 1.0062.

The chart below shows the impact of phase margin and gamma on spur gain. The spur gain does not have a minimum point. As the phase margin is decreased and the gamma value is increased, the spur gain decreases. However, the impact of phase margin and gamma on spur gain is much less than the impact of phase margin and gamma on lock time, so it makes sense to choose the phase margin and gamma value such that lock time is minimized.

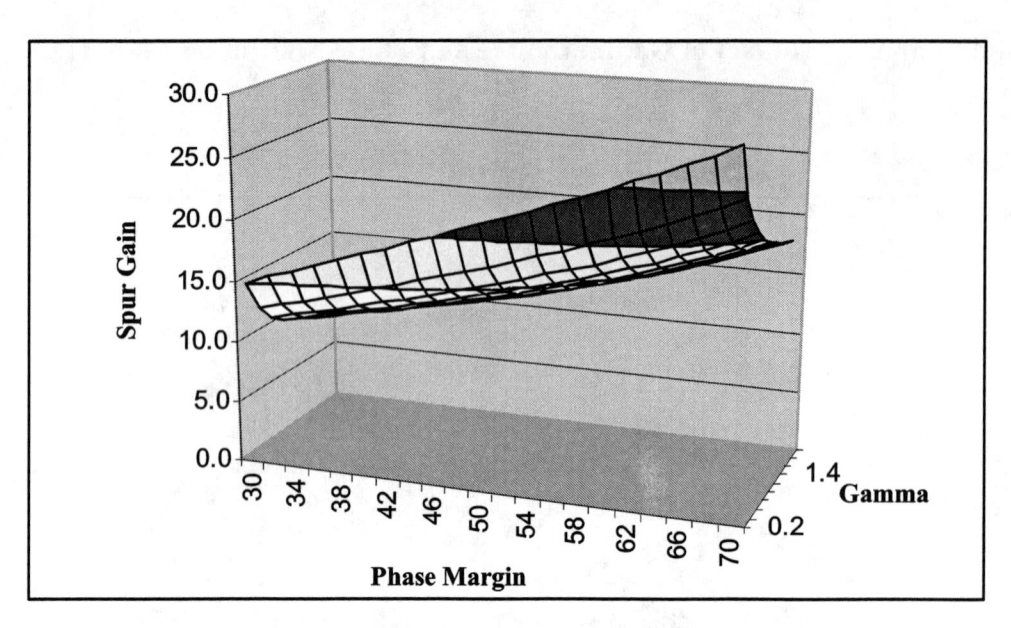

Figure 26.2 *Spur Gain as a Function of Phase Margin and Gamma*

T31	Phi	Gamma	LT	SG
%	Deg	n/a	µS	dB
0	50.8	1.006	246.4	29.9
10	49.8	1.045	243.3	28.7
20	49.0	1.075	240.6	26.6
30	48.2	1.098	238.3	24.8
40	47.8	1.115	236.4	23.7
50	47.4	1.127	235.0	22.9
60	47.1	1.136	233.9	22.4
70	47.0	1.141	233.2	22.0
80	47.0	1.144	232.8	21.9
90	46.7	1.147	232.5	21.7
100	46.8	1.147	232.4	21.7

Table 26.2 *Gamma and Phase Margin Values that Minimize Lock Time*

The above table shows how to choose gamma and the phase margin in order to minimize lock time. These numbers may vary slightly if the frequency jump or frequency tolerance for lock time is changed. One thing that this does not take into consideration is the spur gain. The next simulation does this.

Second Simulation: Optimal Choice of Phase Margin and Gamma to Give the Best Trade-Off Between Lock Time and Spurs

For most designs, it is more realistic to try to minimize lock time while keeping the spur levels constant. Although the loop bandwidth is the most dominant factor, phase margin and the gamma optimization parameter have some impact on spurs. Since lock time and spurs are a trade off, the following table tries to consider both of these by minimizing the following index:

$$Index = 40 \cdot Log \left| \frac{Lock\,Time}{100\,uS} \right| + Spur\,Gain \qquad (26.4)$$

T31 %	Phi Deg	Gamma n/a	LT µS	SG dB
0	49.2	1.024	249.9	29.5
10	46.8	1.081	252.9	27.6
20	44.5	1.144	258.8	24.6
30	43.7	1.168	257.6	22.8
40	43.2	1.184	255.9	21.6
50	42.5	1.203	257.0	20.6
60	42.5	1.204	254.2	20.2
70	42.2	1.212	254.3	19.8
80	42.5	1.207	251.7	19.8
90	42.4	1.209	251.6	19.7
100	42.3	1.211	251.9	19.6

Table 26.3 *Optimal Choices for Phase Margin and Gamma*

The table above is the fundamental result for this chapter. The bottom line is that one should choose *T31* as high as realistically possible for the best lock time and spur performance. Once this parameter is chosen, then the optimal value for phase margin and Gamma can be found from the table. Note that if the frequency jump or tolerance is changed, these numbers change slightly, but this effect is small and can be disregarded for practical purposes. The *T43* ratio was not included because the simulation tool used to generate this table could not model the lock time for this without approximations.

Optimal Phase Margin Choice for Minimum RMS Phase Error

For some applications, lock time is not critical and it is more important to minimize the RMS phase error. For these applications, the loop bandwidth is usually set by a requirement that the PLL needs to be able to pass a certain bandwidth of modulation. The phase margin does have some impact on the RMS phase error. In general, there is some peaking in the closed loop response that happens below the loop bandwidth that is impacted by the phase margin. This peaking has a tendency to dominate the RMS phase error and can be minimized by increasing the phase margin. Simulations were run which tested the impact phase margin on the amount of peaking, the frequency where this peaking occurs, the 0 dB bandwidth, and the RMS phase error. It was found that the loop bandwidth had no impact on the peaking. It did impact the peak frequency and 0 dB bandwidth, it was found that these two parameters were proportional to the loop bandwidth. It was also found that the gamma optimization parameter and pole ratios did have an effect, but this was on the order of 0.1 dB on the peaking and 5% on the peak frequency and 0 dB bandwidth. So for all practical purposes, it is sufficient to study the phase margin alone. The table and figures below show the impact of phase margin, assuming no VCO or resistor noise. For example, if one designs for a phase margin of 40 degrees, there would be 3.95 degrees of peaking, and this peaking would occur at a frequency that is 79% of the loop bandwidth. The 0 dB bandwidth would be 139% of the loop bandwidth, and the RMS phase error would be about 5% more than if one was designing for a phase margin of 50 degrees.

Phase Margin (degrees)	Peaking (dB)	Relative Peak Frequency	Relative 0 dB Banwidth	Relative RMS Phase Error
0	Infinite	1.00	1.41	Infinite
5	21.19	1.00	1.43	1.43
10	15.21	0.99	1.45	1.32
15	11.74	0.98	1.46	1.25
20	9.32	0.96	1.47	1.20
25	7.50	0.94	1.46	1.16
30	6.05	0.90	1.45	1.12
35	4.89	0.85	1.43	1.09
40	3.95	0.79	1.39	1.06
45	3.20	0.71	1.32	1.04
50	2.59	0.63	1.24	1.01
55	2.10	0.55	1.13	0.99
60	1.70	0.47	1.00	0.97
65	1.37	0.39	0.86	0.95
70	1.08	0.32	0.73	0.93
75	0.81	0.25	0.60	0.91
80	0.55	0.18	0.46	0.89
85	0.30	0.11	0.31	0.87
90	0.00	0.00	0.00	0.85

Table 26.4 *Impact of Phase Margin on Various Closed Loop Parameters*

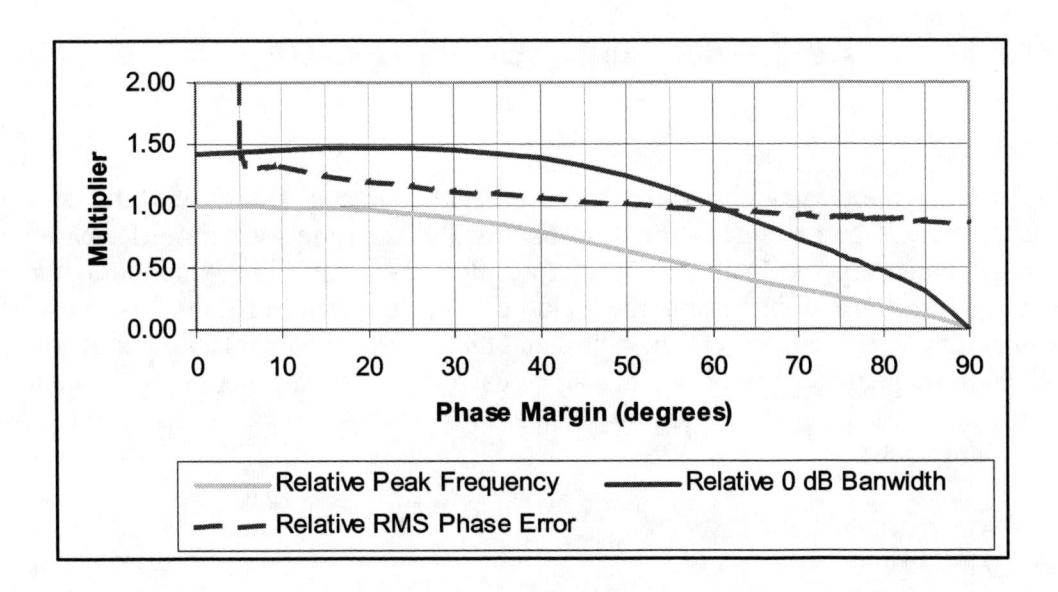

Figure 26.3 *Impact of Phase Margin on Relative Parameters*

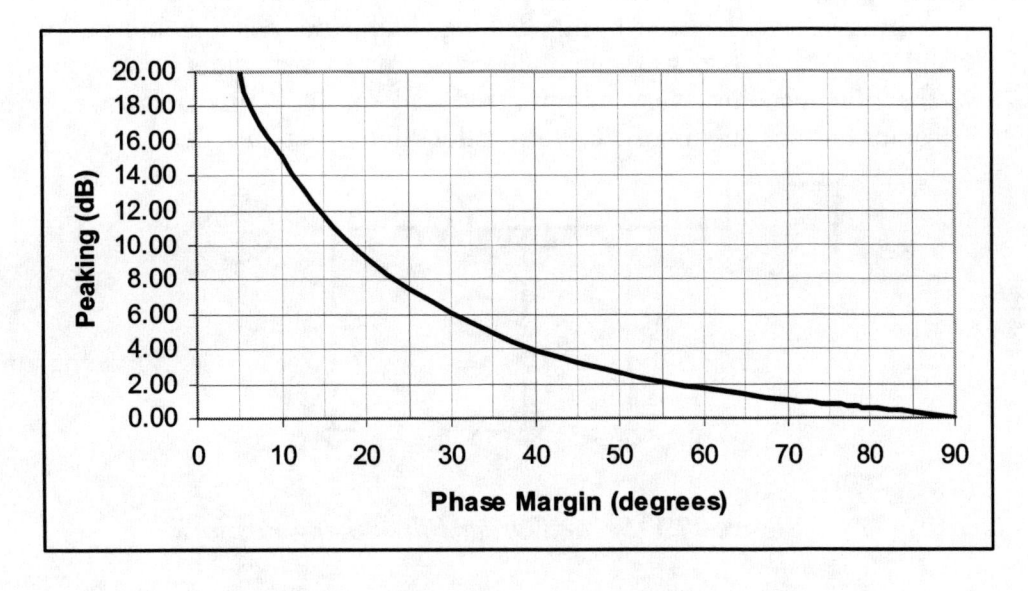

Figure 26.4 *Impact of Phase Margin on Peaking*

Conclusion

The impact of phase margin and gamma optimization parameter have been discussed. For those designs where lock time is critical, it is best to use the simulated results as a guideline. For instance, a second order filter should be chosen with 49.2 degrees of phase margin and a gamma optimization parameter of 1.024. In general, allowing the gamma optimization parameter to be different than one can allow up to a 30% reduction in lock time. For those designs where the lock time is not critical, consider increasing the phase margin to reduce the RMS phase error. This theoretically reduces the peaking a few dB, and this effect tends to be increased when the VCO noise is also considered.

Chapter 27 Using Fastlock and Cycle Slip Reduction

Introduction

In PLL design, there is a classical trade-off between faster switching time and lower reference spurs. If one increases the loop bandwidth, then the lock time decreases at the expense of increasing the spur levels. If one decreases the loop bandwidth, the spurs decrease at the expense of increasing the lock time. The concept of Fastlock is to use a wide loop bandwidth when switching frequencies, and then switch a narrow loop bandwidth when not switching frequencies. Fastlock can also be used in situations where lock time and RMS phase error are traded off, or in situations where lock time and phase noise outside the loop bandwidth are traded off.

Fastlock Description

Fastlock is a feature of some PLLs that allows a wide loop bandwidth to be used for locking frequencies, and a narrower one to be used in the steady state. This can be used to reduce the spur levels, or phase noise outside the loop bandwidth. Fastlock is typically intended for a second order filter. It can be used in higher order loop filter designs, but the pole ratios (*T31*, *T41*, and so on) need to be small, otherwise, when the wider loop bandwidth is switched in, the filter becomes very unoptimized and the lock time increases. For this reason, this chapter focuses only on the use of Fastlock for a second order design.

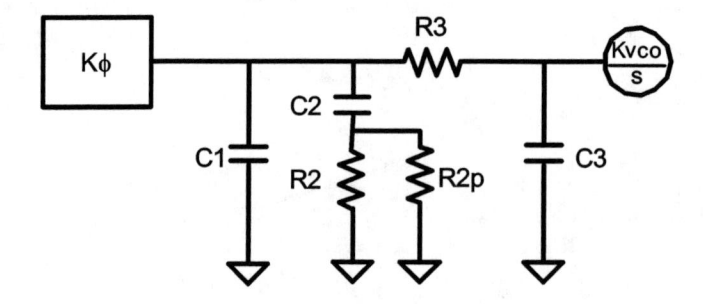

Figure 27.1 *Second Order Filter Using Fastlock*

When the PLL is in the locked state, charge pump gain $K\phi$ is used and resistor $R2p$ is not grounded, therefore having no impact. When the PLL switches frequency, the charge pump gain is increased by a factor of M^2 to $K\phi^*$. Resistor $R2p$ is also switched in parallel with $R2$, making the total resistance $R2^* = R2 \parallel R2p = R2/M$. Recall that the loop filter impedance for the second order filter is given by:

$$Z(s) = \frac{1 + s \cdot C2 \cdot R2}{s \cdot (C1 + C2) \cdot \left(1 + s \cdot \frac{C1 \cdot C2 \cdot R2}{C1 + C2}\right)} = \frac{1 + s \cdot T2}{s \cdot A0 \cdot (1 + s \cdot T1)} \qquad (27.1)$$

$$T2 = R2 \cdot C2 \qquad (27.2)$$

$$T1 = \frac{R2 \cdot C2 \cdot C1}{A0}$$

$$A0 = C1 + C2$$

	Normal Mode	Fastlock Mode
M	$\sqrt{\dfrac{K\phi^*}{K\phi}}$	$\sqrt{\dfrac{K\phi^*}{K\phi}}$
R2p	$\dfrac{R2}{M-1}$	$\dfrac{R2}{M-1}$
Equivalent Resistance, R2*	R2	$\dfrac{R2}{M}$
Charge Pump Gain	$K\phi$	$K\phi^*$
Zero T2	T2	$\dfrac{T2}{M}$
Pole T1	T1	$\dfrac{T1}{M}$
Loop Bandwidth	Fc	M•Fc
Theoretical Lock Time	LT	$\dfrac{LT}{M}$

Table 27.1 *Comparison of Filter Parameters between Normal Mode and Fastlock Mode*

From the above table, one could conclude that if the charge pump was normally 1 mA, and then was switched to 4 mA, *M* would be two and there would be a theoretical 50% improvement in lock time. Another way of thinking about this is that the loop bandwidth could be decreased to half of its original value, thus making a theoretical 12 dB improvement in reference spurs. However, this disregards the fact that there is a glitch when Fastlock is disengaged, and this glitch can be very significant.

The Fastlock Disengagement Glitch

Cause and Behavior of the Glitch

When the Fastlock is disengaged, a frequency glitch is created. This glitch can be caused by parasitic capacitances in the switch that switches out the resistor *R2p*, and also imperfections in charge pump. When the switch is disengaged, a small current is injected into the loop filter. It therefore follows that the size of the glitch is loop filter and PLL specific. One possible way to simulate the glitch is to model the unwanted charge injected into the loop filter as a delta function times a proportionality constant. From this, one can see why the glitch size is greater for an unoptimized filter and inversely proportional to charge pump gain, assuming an optimized loop filter of fixed loop bandwidth. Experimental results show that the ratio, *M*, does not have much impact on this glitch, only the charge pump gain used in the steady state. For instance, if the charge pump gain was 100 µA in normal mode and 800 µA in Fastlock mode, then the glitch caused by disengaging Fastlock would be the same if the current was increased from 100 µA to 1600 µA in Fastlock mode.

The glitch also decreases as the loop bandwidth decreases. This can yield some unanticipated results. For instance, one would think that a loop filter with 2 kHz loop bandwidth using Fastlock would take twice the time to lock as one with a 4 kHz loop bandwidth using Fastlock. However, it could lock faster than this since the Fastlock glitch for the 2 kHz loop filter is less. In other words, the 4 kHz loop bandwidth filter would lock faster than the 2 kHz loop filter, but maybe not twice as fast. Increasing the capacitor *C1* or the pole ratios decrease the glitch, while increasing *C2* makes the glitch slightly larger.

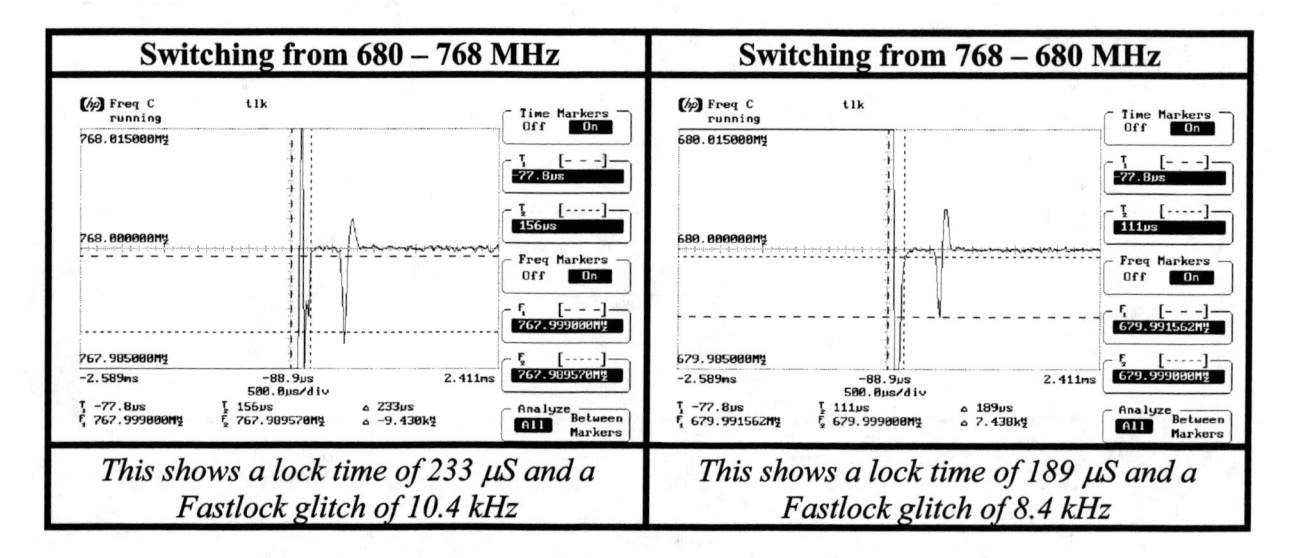

Switching from 680 – 768 MHz	Switching from 768 – 680 MHz
This shows a lock time of 233 µS and a Fastlock glitch of 10.4 kHz	*This shows a lock time of 189 µS and a Fastlock glitch of 8.4 kHz*

Figure 27.2 *Fastlock Disengagement Glitch*

Optimal Timing for Fastlock Disengagement

For optimal lock time, the Fastlock should be disengaged at a time such that the magnitude of this glitch is about the magnitude of the ringing of the PLL transient response. If Fastlock is disengaged too early, then the full benefits of the Fastlock are not realized. If it is disengaged too late, then the settle time for the glitch becomes too large of a proportion of the lock time. Figure 27.2 shows the lock time when the Fastlock glitch is taken into consideration.

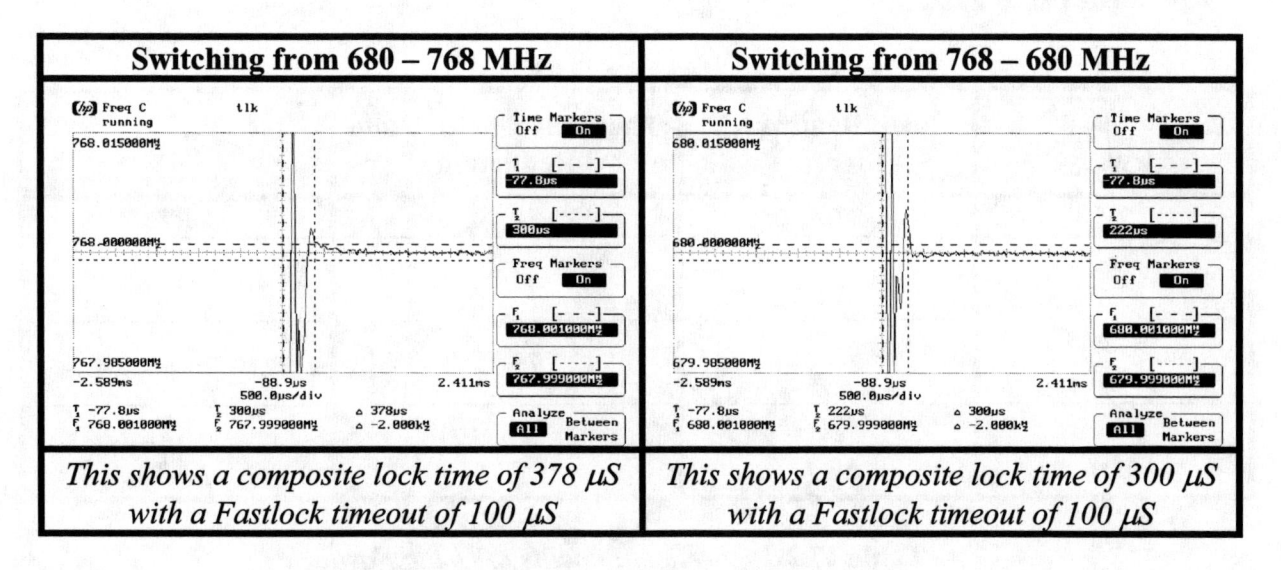

Switching from 680 – 768 MHz	Switching from 768 – 680 MHz
This shows a composite lock time of 378 µS with a Fastlock timeout of 100 µS	*This shows a composite lock time of 300 µS with a Fastlock timeout of 100 µS*

Figure 27.3 *Lock Time Using Optimal Fastlock Timeout of 100 µS*

Disadvantages of Using Fastlock

Increased In-Band Phase Noise

Since Fastlock requires that a higher current is switched in during frequency acquisition, this requires that the PLL is run in less than the highest current mode. Recall from the phase noise chapter that the in-band phase noise is typically better for the higher charge pump gain.

Higher Order Loop Filters

Another disadvantage of using Fastlock is that if one builds a third or higher order filter with much considerable spur attenuation, then it is likely not to work well with Fastlock. Fastlock is most effective for second order loop filters, or higher order filters with small pole ratios.

Benefits of Using Fastlock

$M = \sqrt{\dfrac{K\phi^*}{K\phi}}$	Loop Bandwidth Increase	Theoretical Lock Time Reduction	$R2\text{p}$
2:1	2 X	50 %	$R2$
3:1	3 X	67 %	$\dfrac{R2}{2}$
4:1	4 X	75 %	$\dfrac{R2}{3}$
M:1	M X	$100 \cdot \left(1 - \dfrac{1}{M}\right)$ %	$\dfrac{R2}{M-1}$

Table 27.2 *Theoretical Benefits of Using Fastlock*

The theoretical benefits of using Fastlock presented in the above table should be interpreted as theoretical best-case numbers for expected improvement, since they disregard the glitch caused when disengaging Fastlock. Typically, in the type of Fastlock when the charge pump current is increased from 1X to 4 X (*M=2*), the actual benefit of using Fastlock is typically about 30%. In the type of Fastlock where the charge pump current is increased from 1X to 16X (*M=4*), the actual benefit of using Fastlock is typically closer to a 50% improvement. These typical numbers are based on National Semiconductor's LMX233X and LMX235X PLL families.

Cycle Slip Reduction

When the comparison frequency exceeds about 100 times the loop bandwidth, cycle slipping starts to become a factor in lock time. One technique used by some parts from National Semiconductor involves increasing the charge pump current and decreasing the comparison frequency by the same factor. In this case, all off the loop filter parameters remain the same, but cycle slipping is greatly reduced. This technique works very well in practice. Cycle slip reduction helps to improve the peak time. Normally, the peak time should be about 20% of the total lock time, but if cycle slipping is a problem, it can be the most dominant contributor to lock time. The next several figures show the impact of cycle slip reduction.

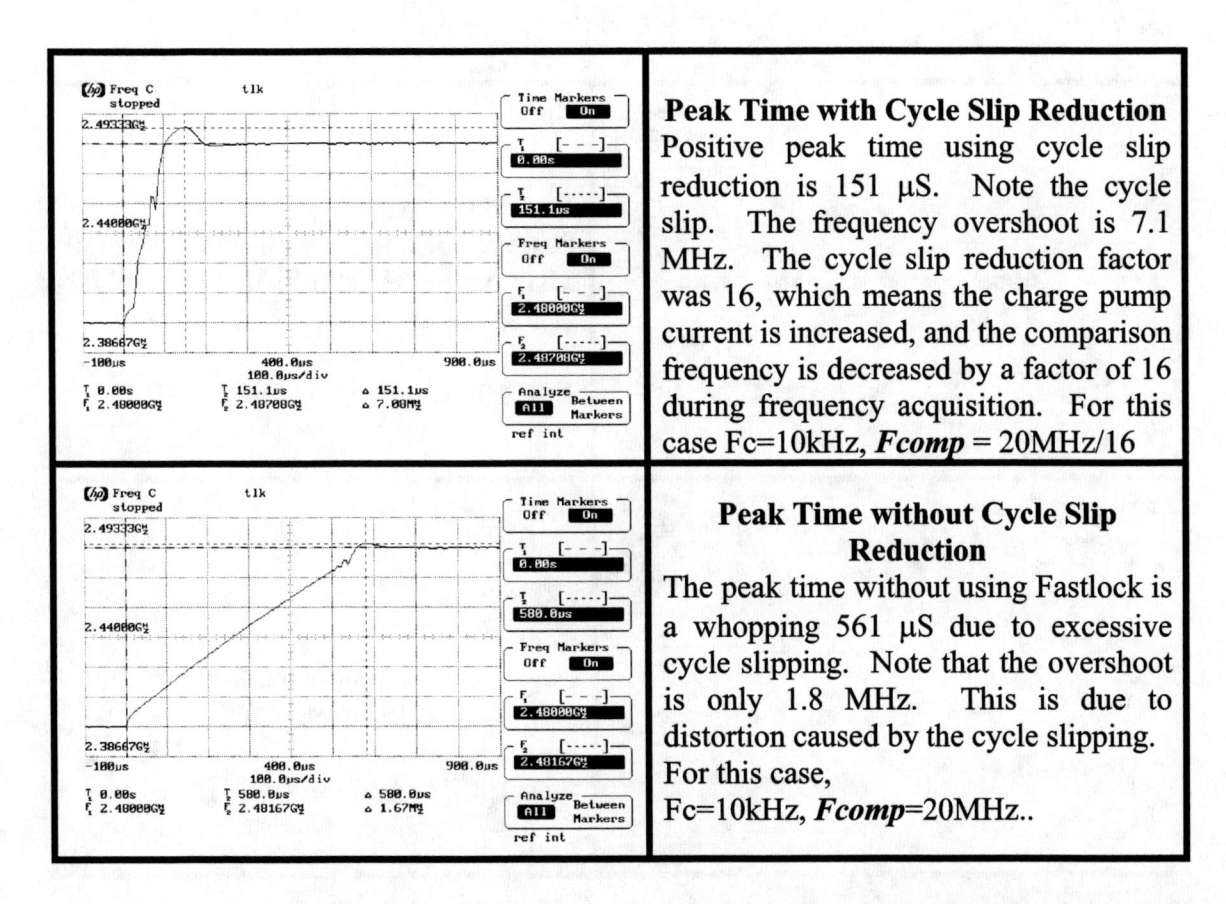

(oscilloscope trace showing frequency response with cycle slip reduction)	**Peak Time with Cycle Slip Reduction** Positive peak time using cycle slip reduction is 151 μS. Note the cycle slip. The frequency overshoot is 7.1 MHz. The cycle slip reduction factor was 16, which means the charge pump current is increased, and the comparison frequency is decreased by a factor of 16 during frequency acquisition. For this case Fc=10kHz, ***Fcomp*** = 20MHz/16
(oscilloscope trace showing frequency response without cycle slip reduction)	**Peak Time without Cycle Slip Reduction** The peak time without using Fastlock is a whopping 561 μS due to excessive cycle slipping. Note that the overshoot is only 1.8 MHz. This is due to distortion caused by the cycle slipping. For this case, Fc=10kHz, ***Fcomp***=20MHz..

Figure 27.4 *Impact of Cycle Slip Reduction On Peak Time*

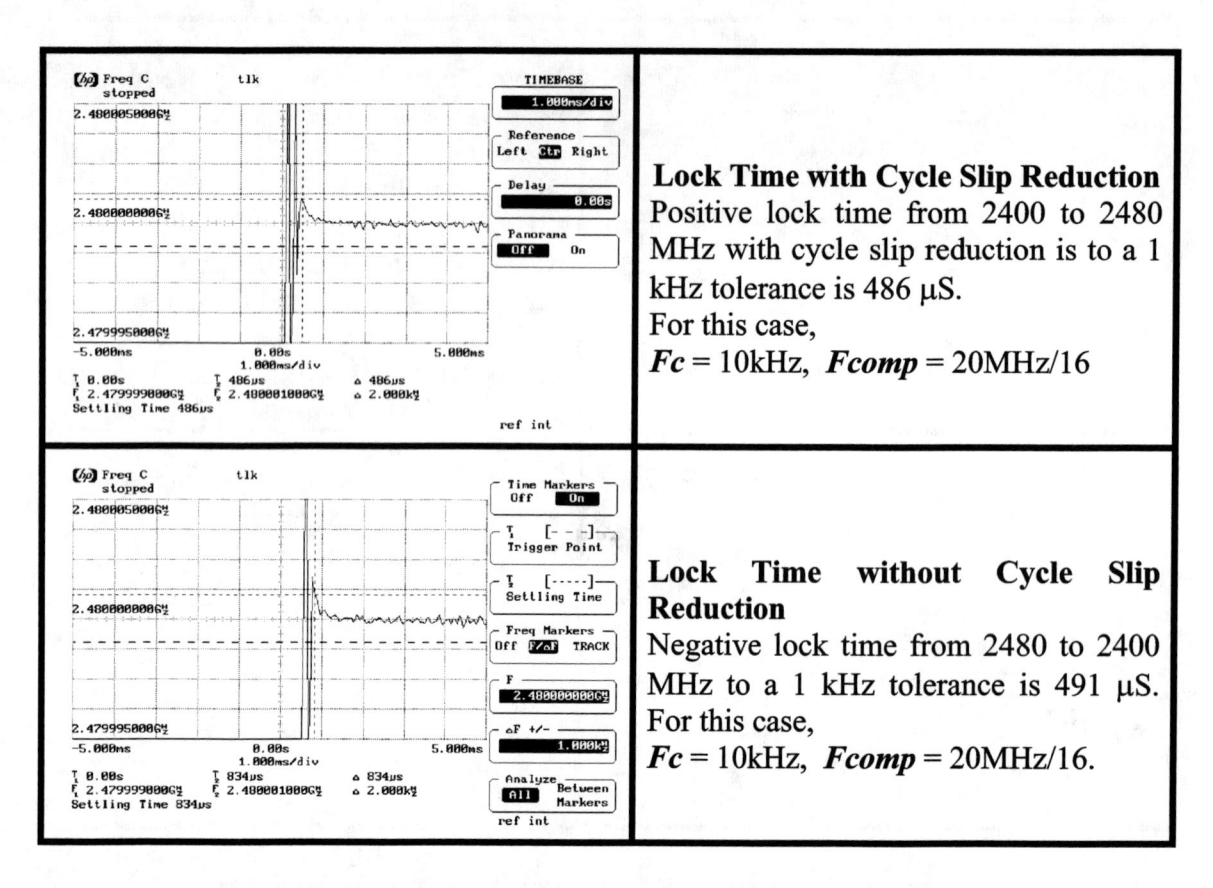

[oscilloscope trace]	**Lock Time with Cycle Slip Reduction** Positive lock time from 2400 to 2480 MHz with cycle slip reduction is to a 1 kHz tolerance is 486 μS. For this case, Fc = 10kHz, $Fcomp$ = 20MHz/16
[oscilloscope trace]	**Lock Time without Cycle Slip Reduction** Negative lock time from 2480 to 2400 MHz to a 1 kHz tolerance is 491 μS. For this case, Fc = 10kHz, $Fcomp$ = 20MHz/16.

Figure 27.5 *Impact of Cycle Slip Reduction on Total Lock Time*

Conclusion

Fastlock is most beneficial in applications where the frequency offset of the most troublesome spur is less than ten times the loop bandwidth. In these situations, higher order filters have little real impact on the spur. As the spur offset frequency becomes farter from the carrier, higher order filters become more practical. An important issue with Fastlock is the glitch created by when it is disengaged. This is application specific, but it can take a significant portion of the lock time.

References

Davis, Craig, et.al.   A Fast Locking Scheme for PLL Frequency Synthesizers.   National Semiconductor AN-1000

Chapter 28 Switched and Multimode Loop Filter Design

Introduction

In some cases, a PLL the same PLL can be used to support multiple modes and frequencies. For instance, some VCOs have a band switch pin that change the frequency band in which they operate. Another example would be a cellular phone that needs a loop filter that supports both the CDMA and AMPS standards. The phase noise, spur, and lock time requirements may be drastically different for these different standards. This chapter explores various types of switched filters.

Loop Gain Constant

The concept used in many switched filters is to keep the loop gain constant.

$$K = K\phi \bullet Kvco/N \qquad (28.1)$$

If the loop gain constant is held the same, and the loop filter components are not changed, then the phase margin, loop bandwidth, gamma optimization factor, and pole ratios will all remain unchanged.

The No Work Switched Filter

The switched filter can be classified by the amount of extra work that is required. This means that it is not necessary to adjust the charge pump gain or the comparison frequency. In some cases, there may be two different VCOs. For the higher frequency VCO, the N value is higher, but the VCO gain might track this reasonably well, so that it is not really necessary to re-design the loop filter. In other cases, it might be that the requirements are lax enough that it is not worth the effort of switching in an additional loop filter. If considering using this approach, the second order loop filter is often a good choice because it is more resistant to changes in the loop gain.

The No Switched Component Filter

In this case, the VCO gain and N value do not track well. Many PLLs have different charge pump current settings. In this case, the charge pump current can sometimes be used. For instance, consider an integer PLL that has a 900 MHz output frequency, but has two modes. The first mode has 30 kHz channel spacing and the second one has 50 kHz channel spacing. So for the mode with 30 kHz channel spacing, if the charge pump current can be roughly adjusted to 5/3 of the value in the other mode, all loop filter characteristics will be preserved.

Using Fastlock for Switched Filters

In this case, the loop gain constant changes too much to ignore. Switching in a Fastlock resistor in parallel with **R2** serves as a quick remedy. In this case, the loop bandwidth may change, but the loop filter stays optimized.

The Full Switched Mode Filter

For this case, a new filter is switched in parallel with the old filter. The most common strategy for using this method is to have one filter with a faster lock time requirement, and one with a slower lock time requirement. For the mode with fast lock time, the other filter is not switched in. For the mode with the slower lock time and better spectral performance, a second loop filter is switched in with components that swamp out the other components.

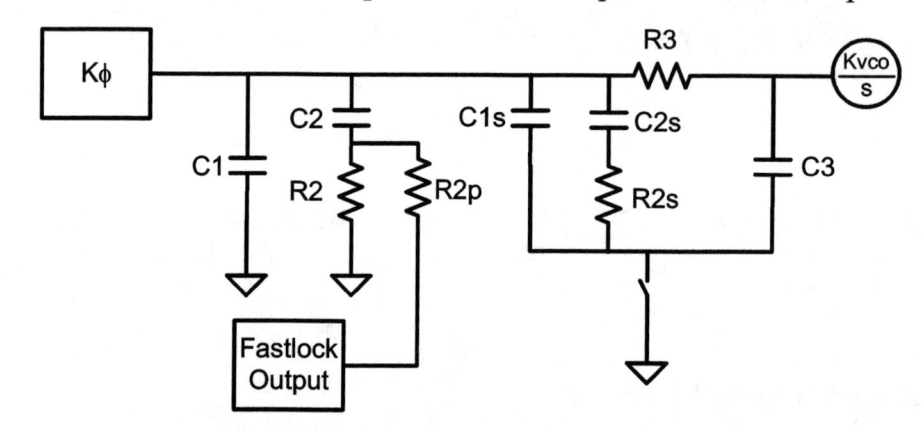

Figure 28.1 *Full Switched Loop Filter*

The strategy with this loop filter design is first to design **C1**, **C2**, **R2**, and **R2**p (Fastlock Resistor) for the mode with fast switching speed. The impact of all the other components is negligible because the switch to ground is off. The components **C3**, **C1s**, **C2s**, and **R2s** add in parallel to **R3** in order to reduce the resistor noise due to this component. In the mode with the more narrow loop bandwidth, the switch to ground is on and **R3** and **C3** form the extra pole for the filter.

Once the filter is designed for the fast mode, then another traditional filter is designed for the slow switching mode. Denote these components with the 'd' suffix. So, **R2d** is the desired component value in slow mode for a non-switched filter. When the switch is grounded, **C1** and **C1s** add together. The transfer function formed by **C2**, **C2s**, **R2**, and **R2s** is as follows:

$$Z(s) = \frac{1 + s \bullet (C2 \bullet R2 + C2s \bullet R2s) + s^2 \bullet C2 \bullet C2s \bullet R2 \bullet R2s}{s \bullet (C2 + C2s) + s^2 \bullet C2 \bullet C2s \bullet (R2 + R2s)} \qquad (28.2)$$

By observing the numerator, it should be apparent that the final transfer function will have a factor of s^2. Because of this, there is no hope of achieving the exact transfer function. Looking at the first term in the denominator, it can be seen that **C2** and **C2s** add to make **C2d**. For **R2**, the middle term should resemble **R2d•C2d**. Now the **C2•R2** makes the calculated value for *R2s* smaller, but the s^2 term would make this smaller. Because these are both second order effects and they roughly cancel out, they can both be neglected. In practice, this approximation seems to work reasonably well. As for *R3s* and *C3s*, all the calculations have been made so far to make the second order part of the loop filter as close as possible, so it makes sense to make these equal to their design target values. Applying all of these concepts, the switched components can be solved for.

$$C1s = C1d - C1 \tag{28.3}$$

$$C2s = C2d - C2 \tag{28.4}$$

$$C3s = C3d \tag{28.5}$$

$$R2s = \frac{R2d \cdot C2d}{C2s} \tag{28.6}$$

$$R3s = R3d \tag{28.7}$$

Note that the way that these switched component values are calculated is by calculating what the equivalent impedance of the loop filter would be with the components switched together and then solving for the switched values. For instance, capacitor *C1* and *C1s* add to get *C1d*. From this, it is easy to solve for *C1s*. Some coarse approximations have been used, so there could definitely be some benefit to tweaking the components manually.

Example of a Full Switched Filter

Symbol	Units	Fast Filter	Ideal Slow Filter	Switched Components for Slow Filter
Fout	MHz	1930-1990	1392 (Fixed Frequency)	
Fcomp	kHz	50	60	
Kφ	mA	1	4	
Kvco	MHz/V	60	30	
N	n/a	39200	23200	
Fc	kHz	10.0	2.0	1.9
φ	Deg.	50.0	50.0	48.9
γ	n/A	1.1	1.1	
T3/T1	%	0	50	
C1, C1d, C1s	nF	0.58494	5.86096	5.27602
C2, C2d, C2s	nF	3.83824	91.17890	87.34066
C3d, C3s	nF		0.75962	0.75962
R2, R2d, R2s	kΩ	23.9181	2.56228	2.67488
R2p	kΩ	23.9181		
R3d, R3s	kΩ		18.57557	18.57557

Conclusion

Switched filters are useful in situations where the loop filter is to be used under two different conditions. In some cases, it is not necessary to switch in additional components. However, if the requirements of the loop filters are much different, then it might be necessary. Also, there can be times when the requirements for two different modes may be different. Usually, this means that there is one mode that has a faster lock time requirement, and another mode that has a more stringent spur requirement.

Chapter 29   Dealing with Real-World Components

Introduction

Much has been said about calculating loop filters. With all this effort going into calculating the theoretical values, it makes some sense to spend a little time discussing how to fit these theoretical values to standard component values and other practical issues when dealing with actual components.

The Basic Method

The most intuitive way to fit standard component values to ideal components is to simply round each component value to the closest standard value. This method is the most intuitive, but does not yield the optimal solution.

The Iterated Calculation Method

This method is based on the basic method and yields better results, but requires a computer. The basic strategy is to vary the parameters such as loop bandwidth and phase margin and simulate each result. Based on the lock time and spur gain of each result, the optimal choice of components can be found this way. This method yields the best results.

The Advanced Rounding Method for a Passive Loop Filter

The object of this method is to keep the loop filter coefficients as close to the theoretical values as possible. This analysis will be limited to a passive filter to simplify matters. The key to this technique involves understanding what components have the dominant on which parameter. The steps for rounding are as follows:

Step 1: Choose Capacitor C2 as Close as Possible

The loop bandwidth is perhaps the most dominant factor, and the largest influence on this is the sum of the loop filter capacitors. Because capacitor $C2$ is the largest capacitor, the first step is to choose $C2$ as close as possible.

Step 2: Choose R2 to Make R2•C2 as Close to Design Value as Possible

The time constant, $T2$ has a very large impact on phase margin and gamma optimization factor. Because $T2$ is the product of these components, $R2$ should be chosen to preserve this time constant. So if the actual standard component value for $C2$ is 3% lower than the theoretical value, then the most desirable scenario would be for $R2$ to be 3% higher.

Step 3: Choose C1 to Make C2/C1 as Close to Design Value as Possible

For a second order filter, it can be shown that the ratio of *C2/C1* has the largest impact on phase margin, after loop bandwidth and *T2*. Because *C2* is know, now *C1* can be calculated as well.

Step 4: Choose C3 and C4 as Close to Design Value as Possible

The next step is to get the poles *T3* and *T4* as close as possible to the design values. *T3* is most dominated by the product of *R3* and *C3*, and *T4* is most dominated by the product of *R4* and *C4*. Resistors are easier to stock and it is more likely that these values will be available. With capacitors, it might be the case that not all values are available. This can become the case for the larger values, or if one wants to use a particular type of capacitor, like film.

Step 4: Choose R3 and R4 to Make R3•C3 and R4•C4 as Close to Design Value as Possible

In order to best match the time constants *T3* and *T4*, choose *R3* such that the product of *R3•C3* is as close to the design value as possible and that *R4•C4* is as close to design value as possible.

A Component Rounding Example

Parameter	Units	Ideal Components	Basic Method	Advanced Method
Kφ	mA	4		
Kvco	MHz/V	20		
N	n/a	4500		
Loop Bandwidth	kHz	5	5.21	4.59
Phase Margin	Degrees	45	45.2	44.8
T3/T1	%	50	50.03	52.1
Gamma Optimization Factor	n/a	1.18	1.59	0.92
Lock Time (889-915 MHz to 1 kHz)	µS	442.1	622	610
Spur Gain @ 200 kHz	dB	4.04	3.39	3.36
C1	nF	9.388859	10	10
C2	nF	112.220087	120	
C3	nF	1.240375		
R2	kΩ	0.763858	0.82	0.68
R3	kΩ	5.335533	5.6	

Table 29.1 *Example of Component Rounding with a Third Order Passive Filter*

The above table shows an example of rounding components to the nearest 10% value. In this case, even though components do not come with a 10% tolerance, it is a very common practice to order 5% components and stock every other value. The effect of this is the same as having standard 10% values. These values are a power of ten multiplied by one of the following values: 1.0, 1.2, 1.5, 1.8, 2.2, 2.7, 3.3, 3.9, 4.7, 5.6, 6.8, or 8.2. For the case of the basic method, the components were simply rounded to the nearest value.

The advanced method requires a little bit more work. The first step is to round **C2**. In this case, it rounds to 120 nF. Now to calculate **R2**, the adjusted target value is 0.713 kΩ, which rounds to 0.68 kΩ. The next step is to choose **C1** as close as possible to the adjusted target value of 10.05 nF, which works out to 10 nF. All the other values for the advanced method work out the same as in the basic method. The advanced method slightly outperforms the basic method in this case for both lock time and spur gain.

Dealing With Capacitor Dielectrics

When it comes to loop filters, there is not much more to resistors than their value. Power dissipation is not an issue, so there is no advantage to choosing a larger footprint. However, with capacitors, there are some issues. There are situations where there is a trade-off between the physical size of the capacitor and the quality of the dielectric. Dielectric type has no noticeable impact on spurs or phase noise, but can have a very large impact on lock time in some applications. From practical experience, capacitors with a Film dielectric or NP0/C0G dielectric perform very close to how a theoretical capacitor would do. However, these capacitors may require a larger footprint or simply not be available for larger capacitor values. X7R dielectric can cause increases in lock time from 0-500%. In general, it seems that designs with higher comparison frequencies are less susceptible to capacitor dielectrics. High comparison frequencies would be considered several Megahertz. Tantalum capacitors are not recommended for loop filters. By far, the capacitor for which the dielectric is most important is capacitor *C2*, which is also the largest capacitor. Capacitor *C1* has an impact, but not nearly as much as *C2*.

It seems that the theoretical performance parameter for the capacitor is dielectric absorption. Dielectric absorption is measured by applying a voltage to a capacitor, then shorting the capacitor, then removing the short. A residual voltage develops across the capacitor and is related to this dielectric absorption. The impact of this is to draw out the final fine frequency settling time of the PLL.

What is a nanofarad (nF)?

The nanofarad is 10^{-9} Farads. This is no surprise, but the use of this unit is controversial. There are some that feel that one should express all capacitance values in terms of picofarads (pF) and micorfarads (µF). A search of the National Institutes of Standards and Technology (www.nist.gov) will reveal that nF is used in various places. Since this book is more focused on new ideas and not just following the status quo, this book boldly uses the unit of nF in the face of harsh criticism.

Conclusion

The impacts of standard components should be considered. The first consideration is standard component values. Although the easiest approach is to simply round off the theoretical component values to their standard values, performance can often be enhanced by using the advanced method of component rounding. Loop filter resistors tend to behave just as they should, but capacitors in the loop filter can often give undesired effects, especially the bigger ones. Note that this chapter is only concerned with how these components behave in a loop filter. There are many more non-ideal effects that resistors and capacitors can have at high frequency as well, but these do not impact loop filter performance.

Additional Topics

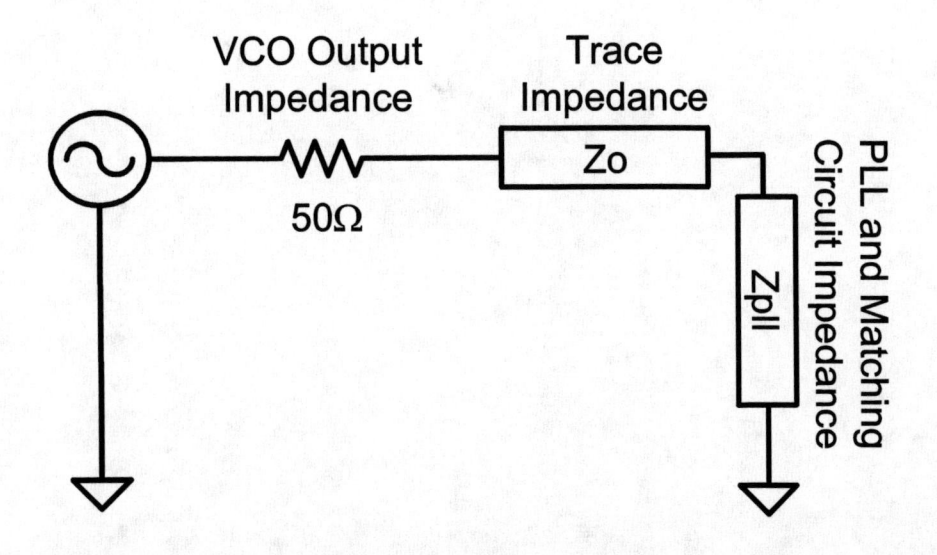

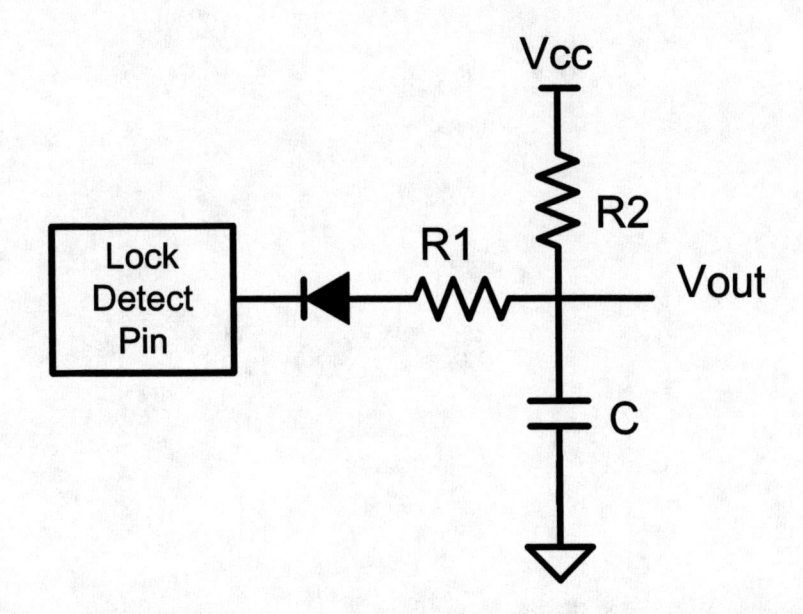

PLL Performance, Simulation, and Design © 2003, Third Edition

Chapter 30 Lock Detect Circuit Construction and Analysis

Introduction

Although many newer PLLs have a lock detect pin that give a logic level output to indicate whether or not the PLL is in lock, there are still many PLLs, including the LMX233X series from National Semiconductor, that do not put out a logic level signal to indicate whether or not the part is in lock; external circuitry is necessary in order to make meaningful sense of the signal. This chapter discusses the design and simulation of such a circuit.

Using the Analog Lock Detect Pin

The state of analog lock detect pin is high when the charge pump is off and low when the charge pump turns on. When viewed with an oscilloscope, one can observe narrow negative pulses that occur when the charge pump turns on. When the PLL is in the locked state, these pulses are on the order of 25-70 nS in width; however, this number can vary based on the VCO gain, loop filter transfer equations, phase detector gain, and other factors, although it should be constant for a given application. For some PLLs, the output is open drain and requires a pull-up resistor to see the pulses.

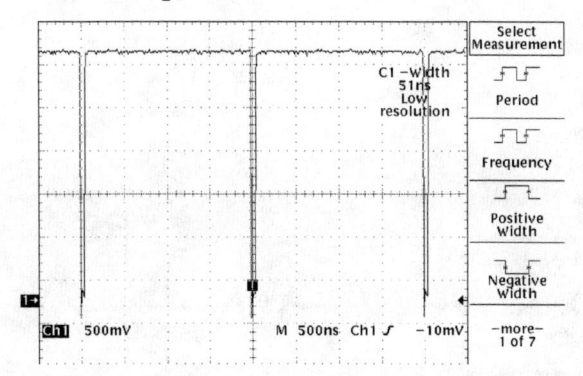

Figure 30.1 *Lock Detect Pin Output for a PLL in the Locked State*

When the PLL is not in the locked state, the average width of these pulses changes. The information concerning the PLL in or out of the locked state is in no individual pulse, but rather in the average pulse width. If the VCO kept on but disconnected from the charge pump, then the signal from the lock detect pin will have a duty cycle that oscillates between a low and high duty cycle. However, this is unrealistic, since the PLL tries to keep the VCO in phase. When the VCO is connected to the PLL, but is off frequency, the pulse width is much more predictable and closer to being constant. The pulses are sort of triangular due to the turn on times of transistors and other effects. For the sake of simplicity and simplifying calculations, they will be treated as rectangular. For a ballpark estimate of how much the average width of the pulses will change and a rough idea on how sensitive the circuit is, the average change in the width of the pulses at any given time could be approximated by the difference in the periods of the N counter and the R counter. This result was discussed in a previous chapter concerning the performance of the phase detector. In other words,

$$\text{Change in Average Pulse Width} = Tlow - Tloc = \frac{1}{Fcomp} - \frac{N}{Fout} \qquad (30.1)$$

Lock Detect Circuit Construction

The basic strategy for the type of lock detect circuit described in this chapter is to integrate over some number of reference periods in order to accumulate some DC value which can then be compared to a threshold value. This comparison can be made with a comparator or transistor. In cases where only a gross lock detect is needed, the lock detect circuit output can be sent directly to the input logic gate, provided the difference in the voltage level produced between the in lock and out of lock conditions is large enough to be recognized as a high or low. Some microprocessors also have A/D input pins that can also be used for this function.

Since the average DC contributions of the pulses are so small relative to the rest of the time, it may be necessary to use unbalanced time constants to maximize sensitivity. The recommended circuit is shown in Figure 30.2 . Note that there are some PLLs in which the lock detect output is open drain, which eliminates the need for the diode. There are still other PLLs with digital lock detect, that eliminate the need for a lock detect circuit entirely.

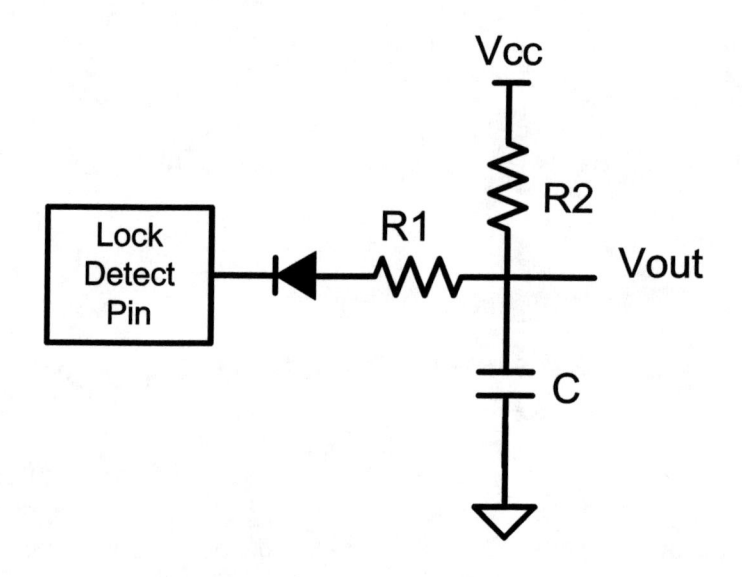

Figure 30.2 *Lock Detect Circuit*

Theoretical Operation of the Lock Detect Circuit

Consider the event when the lock detect pin first goes to its low voltage, V_{OL}. The voltage drop across the diode is V_D. The diode will conduct, and if $R2 \gg R1$ then the following holds:

$$V_{out} = -R1 \cdot C \cdot \frac{dV_{out}}{dt} + V_L \qquad (30.2)$$

$$V_L = V_D + V_{OL} \qquad (30.3)$$

What is really of interest is how much does the voltage V_{out} change during the period that the lock detect pin is low. To simplify the mathematics, it is easiest to discretize the problem. The size of the discrete time step is T_L, which is the time which the lock detect pin stays low. The following definitions can be used to convert the differential equation into a difference equation:

$$V_n = V_{out}(0) \qquad (30.4)$$

$$V_{n+1} = V_{out}(T_L) \qquad (30.5)$$

The above differential equation has the following solution:

$$V_{n+1} = V_L + (V_n - V_L) \cdot \beta \qquad (30.6)$$

$$\beta = e^{\frac{-T_L}{R1 \cdot C}} \qquad (30.7)$$

When the lock detect output goes high, then the diode will not conduct, and the capacitor will charge through the resistor $R2$. In an analogous way that was done for the case of the lock detect pin state being low, the results can also be derived for the case when the lock detect pin is high. In this case, T_H represents the time period that the lock detect pin stays high.

$$V_{n+1} = Vcc + (V_n - Vcc) \cdot \alpha \qquad (30.8)$$

$$\alpha = e^{\frac{-T_H}{R2 \cdot C}} \qquad (30.9)$$

Now if one considers the two cases for V_n, then a general expression can be written for V_n. For sufficiently large n, the series will alternate between two steady state values. Call these two values V_{High} and V_{Low}. These values can be solved for by realizing that the initial voltage when the lock detect pin just goes low will be V_{High} and the final voltage will be V_{Low}. Also, the initial voltage when the lock detect pin just goes high will be V_{Low} and the final voltage will be V_{High}. This creates a system of two equations and two unknowns.

$$V_{Low} = V_L + (V_{High} - V_L) \bullet \beta \qquad (30.10)$$

$$V_{High} = Vcc + (V_{Low} - Vcc) \bullet \alpha \qquad (30.11)$$

This system of equations has the following solution:

$$V_{Low} = Vcc + \frac{(1-\beta) \bullet (V_L - Vcc)}{1 - \alpha \bullet \beta} \qquad (30.12)$$

$$V_{High} = V_L + \frac{(1-\alpha) \bullet (Vcc - V_L)}{1 - \alpha \bullet \beta} \qquad (30.13)$$

Lock Detect Circuit Design

The above expressions for V_{Low} and V_{High} show what two values the voltage will oscillate between in the locked condition, once the component values are known. These equations can be worked backwards to solve for component values as well. For design of the circuit, the following information is needed.

T_{lock} The width of the pulses in the locked condition. This should be around 25 nS for the 4X current mode and 50 nS for the 1X current mode.

T_{switch} The width of the LD pulses that are to be detected.

V_{high} The "trip point". In the unlocked condition, the maximum voltage output would be V_{high}. In the locked condition, the voltage output should be higher

Ripple $V_{high} - V_{low}$. This should be a couple hundred millivolts. Designing for too much ripple can cause a noisy circuit, while designing for too little will cause the circuit to take longer to settle to its final values of V_{low} and V_{high}

Using the expressions for V_{high} and V_{low}, the following equations can be derived.

$$\alpha^2 \cdot A + \alpha \cdot B + C = 0 \qquad (30.14)$$

$$A = K \cdot (V_L - V_{High})$$
$$B = Vcc - V_{High} - K \cdot V_L + K \cdot V_{High}$$
$$C = V_{High} - Vcc$$
$$K = \frac{Vcc - V_{Low}}{V_{High} - V_{Low}}$$

α and β can be solved for as follows:

$$\alpha = \frac{-B + \sqrt{B^2 - 4 \cdot A \cdot C}}{2 \cdot A} \qquad (30.15)$$

$$\beta = 1 + (\alpha - 1) \cdot K$$

Finally, the components can be solved for. To do so, the capacitor, C, can be chosen arbitrarily. Once C is known, the other components can also be found.

$$R1 = \frac{-T_L}{C \cdot \ln \beta} \qquad (30.16)$$

$$R2 = R1 \cdot \frac{\ln \alpha}{\ln \beta} \cdot \frac{T_H}{T_L}$$

Voltages	Volts	Times	ns	Design Specification	Volts
V_D	0.7	T_L	55	$V_{High\ (unlocked)}$	2.1
V_{OL}	0.5	T_H	1600	Ripple Voltage	0.1
Vcc	4.1				
Constants		Components		Calculated Values	
K	2.3333	Choose C1	220 pF	R1	2.12 $K\Omega$
A	-2.1			R2	149.1 $K\Omega$
C	-2			$V_{Low\ (unlocked)}$	2 Volts
α	0.9524				
β	0.8889				

Table 30.1 *Typical Lock Detect Circuit Design*

Simulation

Note that after the design is done, it is necessary to assure that the lowest voltage in the locked state $V_{Low\ (locked)}$ is higher than the highest voltage unlocked condition $V_{High\ (unlocked)}$. In Table 30.2, the circuit designed in Table 30.1 is simulated. The simulation shows that in ten reference cycles, the circuit gets reasonably close to its final steady state values. When the PLL is in lock, the lock detect circuit output voltage will not go below 2.54 Volts; in the unlocked state, the output voltage will not go above 2.10 Volts. This may not seem like much voltage difference, but this is because this circuit is extremely sensitive. If one was to use a pulse width of 100 ns out of lock, then this voltage difference would be much greater.

Table 30.2 shows the simulation of a lock detect circuit. It is necessary to include a lot of margin for error, since it is very difficult to get an accurate idea of the width of the negative pulses from the lock detect pin. It was also assumed that these pulses were square and of constant period, which may be a rough assumption. Furthermore, as shown below, it does take time for the system to settle down to its final state.

Par.	Volts	Components			Times	nS	Const.	Volts	Locked Parameters		
V_D	0.7	C	220	pF	T_L	55	α	0.9524	T_{lock}	25	ns
V_{OL}	0.5	R1	2.1	KΩ	T_H	1600	β	0.8888	$β_{lock}$	0.9478	V
Vcc	2.1	R2	149	KΩ							
Vstart	4.5										

Iter.	Vhigh	Vlow		Iter.	Vhigh	Vlow		Steady State Parameters		
0	2.5000	2.3554	Volts	8	2.2051	2.0933	Volts	$V_{High\ (unlocked)}$	2.0996	Volts
1	2.4385	2.3007	Volts	9	2.1889	2.0789	Volts	$V_{Low\ (unlocked)}$	1.9995	Volts
2	2.3864	2.2545	Volts	10	2.1751	2.0667	Volts	Ripple	0.1001	Volts
3	2.3424	2.2153	Volts	11	2.1635	2.0564	Volts	$V_{Low\ (locked)}$	2.5451	Volts
4	2.3051	2.1822	Volts	12	2.1537	2.0476	Volts			
5	2.2735	2.1541	Volts	13	2.1454	2.0402	Volts			
6	2.2468	2.1304	Volts	14	2.1384	2.0340	Volts			
7	2.2242	2.1103	Volts	15	2.1324	2.0287	Volts			

Table 30.2 *Typical Lock Detect Circuit Simulation*

Conclusion

This chapter investigated some of the concepts behind a lock detect circuit design. It is necessary for the designer to have some idea how much the width of the lock detect pulses are changing between the locked and unlocked condition. For both of these situations, T_L was used to represent the width of these lock detect pulses. It is here that it may be necessary to make some gross estimates. Once T_L is known, then the voltage levels of the circuit in the locked and unlocked condition can be calculated. Since there is ripple on this voltage, the minimum voltage level in the locked state should be greater than the maximum voltage level in the high state. From this pulse width, the components can be calculated. Note that there is a trade-off between the sensitivity of the circuit and the time it takes the circuit to respond, as seen in the simulation. Although ripple is undesirable, some ripple must be tolerated in order for the circuit to have sufficient sensitivity. One possible variation of the circuit is to design for a high amount of ripple and then add additional low pass filtering stages afterwards. There is also a specific choice of time constants for theoretical optimum sensitivity. However, assumptions need to be made about the pulse width and the pulse shape, there will be some tinkering left to the lock detect circuit designer.

Chapter 31   Impedance Matching Issues and Techniques for PLLs

Introduction

This chapter is devoted to matching the VCO output to the PLL input. In most cases, the VCO has a 50 Ω output impedance. However, the PLL input impedance is usually not purely real and not 50 Ω. This can be the cause of many strange problems and a source of tremendous confusion. If the PLL impedance differs greatly from the trace impedance, then power will be reflected back towards the VCO, and significant power will be lost. Furthermore, if the PLL input impedance is not 50 Ω, then this can also cause misinterpretations of the VCO output power level, since it is typically specified for a 50 Ω load. This chapter discusses some of the issues and problems that can arise because of the PLL input impedance not being 50 Ω, and also provides some general matching techniques.

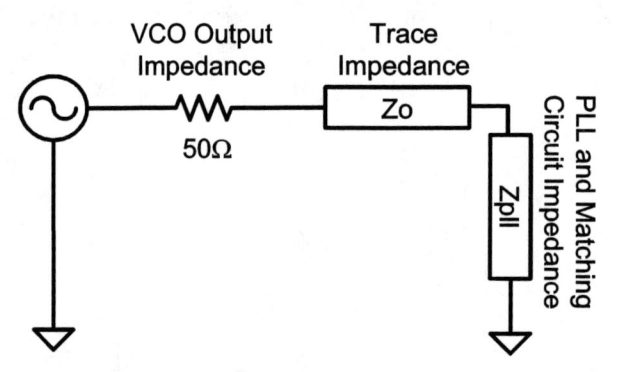

Figure 31.1 *Circuit Between VCO and PLL*

Calculation of the Trace Impedance

The characteristic impedance of the trace between the PLL and the VCO is determined by the width of the trace, W, the height of the trace above the ground plane, H, and the relative dielectric constant, ε_r, of the material used for the PCB board. The reader should be careful to not confuse the characteristic impedance of a microstrip line with the input impedance of the PLL or the output impedance of the VCO; these things are all different.

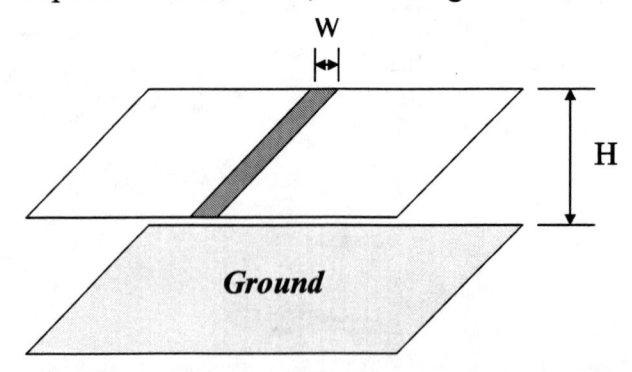

Figure 31.2 *Calculation of Trace Impedance*

The precise calculation of the trace impedance is rather involved, as is the solution. It is a reasonable approximation to say that the trace impedance is independent of frequency, and it can be approximately calculated with the following formula from the first reference:

$$Zo \equiv \sqrt{\frac{L}{C}} = \frac{87}{\sqrt{\varepsilon_r + 1.41}} \bullet ln\left(7.5 \bullet \frac{H}{W}\right) \qquad (31.1)$$

In this formula, **L** represents the inductance per unit length and **C** represents the capacitance per unit length. This formula can also be rearranged in order to determine what ratio of height to width is necessary to produce the desired impedance:

$$\frac{H}{W} = \frac{e^{\frac{Zo \bullet \sqrt{\varepsilon_r + 1.41}}{87}}}{7.5} \qquad (31.2)$$

FR4 is a commonly used material to make PCB boards which has the property that $\varepsilon_r = 4$. This implies that the ratio of the height to the width is about 0.5 for a 50 Ω trace. In other words, if the thickness from the top layer to the ground plane is 31 mils (thousandths of an inch), then the width of the trace should be 62 mils. There are many online calculators for microstrip impedance, such as the first reference presented.

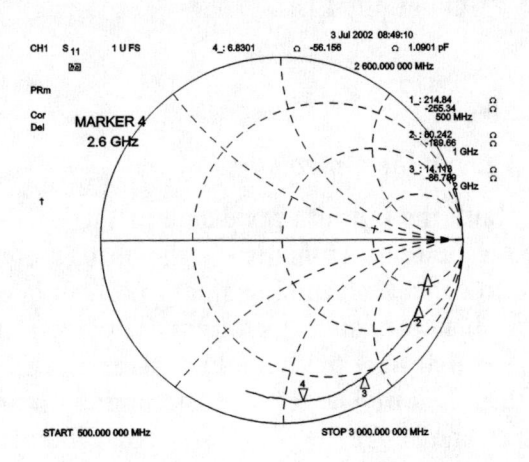

Figure 31.3 *Smith Chart for Typical Input Impedance for a PLL*

Problems with Having the Load Unmatched to the PCB Trace

Throughout this chapter, the trace impedance will be assumed to be 50 Ω, but the PLL impedance will be assumed to be something different. Note from Smith Chart in Figure 31.3 that the input impedance of the PLL is far from 50 Ω and is also frequency dependent. It is very common for PLLs to have an input impedance with a negative imaginary part (i.e. capacitive). In cases where the signal frequency is low, few problems arise. However, for signals in the GHz range, impedance matching problems are common. In the GHz range, a trace of more than a couple centimeters can cause problems if the PLL impedance is poorly matched to the trace impedance. This typically causes a loss of power and can agitate sensitivity problems in the PLL. Also, since VCOs also put out harmonics, it could cause the prescaler to miscount on a higher harmonic of the VCO if the mismatch is severe enough. In most cases, it is not necessary to use any matching network at all. One way to determine how well the PLL is matched to a 50 Ω line is to calculate the reflection coefficient.

$$\rho = \sqrt{\frac{(Ra-Ro)^2 + Xa^2}{(Ra+Ro)^2 + Xa^2}} = \sqrt{\frac{reflected\ power}{transferred\ power}} \qquad (31.3)$$

The above formula assumes the impedance of the transmission line is Ro, and the impedance of the PLL is $Ra + j \bullet Xa$. If the reflection coefficient is one, then no power is transferred to the PLL, if it is zero, all the power is transferred to the PLL. If the reflection coefficient gets too large, then this could cause problems. These problems are most pronounced when there is a long trace between the VCO and the PLL.

Impedance Matching Strategies

Eliminating the Imaginary Part of the Impedance

Without loss of generality, both the output impedance of the VCO and the input impedance of the PLL can be assumed to be real. If this is not the case, it can be made so by putting a series capacitor or inductor to cancel out the imaginary part. It is common for PLLs to have a negative reactance; and in this case, an inductor can be placed in series to cancel this out. The problem with inductors is that they tend to add cost, and this is not necessary unless the negative reactance of the PLL is fairly large. An alternative approach is to use a capacitor beyond its self-resonant frequency. At frequencies beyond the self-resonant frequency, the capacitor actually acts like an inductor and adds a positive series reactance. Because a capacitor is required anyways as a DC blocking component, this approach is very economical.

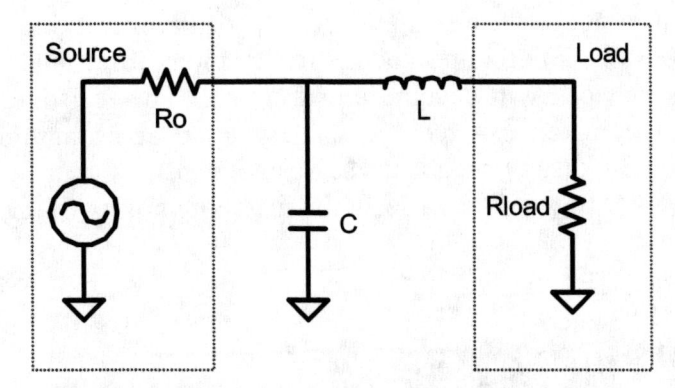

Exactly Matching any Two Real Loads at a Fixed Frequency

Figure 31.4 *Typical Impedance Matching Circuit*

For this type of match, the frequency must be specified. Note also that this assumes that the load resistance is greater than the source resistance. If this is not the case, then the inductor L, needs to be moved to the left hand side of capacitor *C*, instead of the right hand side and the values for the load and source resistance need to be switched. The matching circuit is designed so that both the load and source see a matching impedance. This yields a system of two equations and two unknowns that can be calculated *L* and *C*. In the case that the load has a negative reactance and also has less resistance than the source, it is convenient to compensate for the negative reactance by making the inductor, *L*, bigger by the appropriate amount.

$$\frac{Ro}{1+s \cdot C \cdot Ro} + s \cdot L = Rload \tag{31.4}$$

$$\frac{s \cdot L + Rload}{s^2 \cdot L \cdot C + s \cdot Rload \cdot C + 1} = Ro \tag{31.5}$$

Solving these simultaneous equations yields the following:

$$C = \frac{\sqrt{\frac{Ro}{Rload} - 1}}{\omega \cdot Ro} \tag{31.6}$$

$$L = C \cdot Ro \cdot Rload \tag{31.7}$$

The Resistive Pad

Although the method in the previous section can match any load to any source exactly, it is often not used because inductors are expensive. Also this method is only designed for a fixed frequency and PLL input impedance. If the input impedance of the load varies drastically, then this network will become unoptimized. The resistive pad is a method of matching that does not match exactly, but is very good at accounting for variations in impedance. The biggest disadvantage of the resistive pad is that VCO power must be sacrificed. As more VCO power is sacrificed, the matching ability of the pad increases.

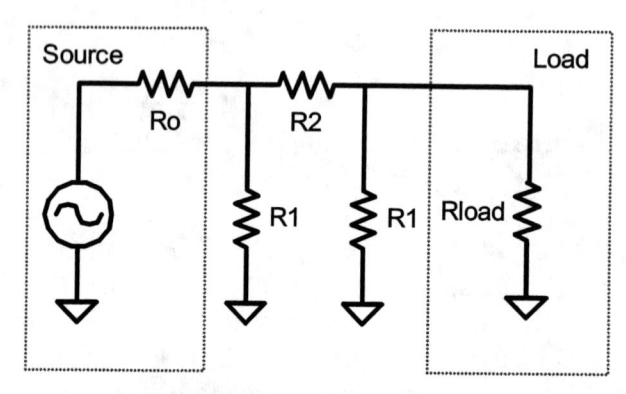

Figure 31.5 *Typical Resistive Pad*

For the resistive pad, the attenuation of the pad is specified, and it is designed assuming that both the source and load impedance are equal to *Ro*, usually 50 Ω. The resistor values satisfy the following equations.

$$R1 \| (R2 + R1 \| Ro) = Ro \qquad (31.8)$$

$$\frac{(R1 \| Ro) \cdot R1}{R1 + R2 + R1 \| Ro} = 10^{\frac{Atten}{20}} = K \qquad (31.9)$$

In these equations, *Ro* is the source impedance, *Atten* is the attenuation of the pad, and $x \| y$ is used to denote the parallel combination of two components, *x* and *y*. The components *R1* and *R2* can be calculated as follows:

$$R1 = Ro \cdot \frac{K+1}{K-1} \qquad (31.10)$$

$$R2 = \frac{2 \cdot Ro \cdot R1}{R1^2 - Ro^2} \qquad (31.11)$$

Adjusting the Trace Width to Match the PLL Input Impedance and Keeping Traces Short

Regardless of whether a resistive pad or LC matching network is used, the idea was to make the load impedance look the same as the source impedance. If these impedances are matched, then the trace impedance can be made equal to these impedances, and there will theoretically be no undesired transmission line effects, such as standing waves. Another matching strategy is to match the trace impedance to the PLL input impedance, instead of the VCO output impedance. The matching of the trace impedance to the PLL impedance is much more important than the matching of the trace impedance to VCO output impedance. Also, if the trace is short ($1/10^{th}$ of a wavelength or less), then transmission line effects are much less likely to be present.

Real World Component Effects at High Frequencies

In the design of impedance matching networks for high frequency, one should be aware of some of the characteristics of real-world components. Some of the relevant behaviors of resistors and capacitors are discussed below.

Capacitors

One classical problem is choosing what capacitor is optimal to filter out a given frequency. Theoretically, the larger the capacitor, the more the filtering. However, capacitors have an equivalent series resistance (ESR) which limits the minimum impedance at high frequencies. At higher frequencies the impedance due to the ESR can be larger than the impedance due to the capacitance value. In general, larger capacitor values tend to lead to a bigger ESR. Although the ESR is component specific, a quick estimate for this would be on the order of $1\ \Omega$.

Figure 31.6 *High Frequency Capacitor Model*

Another phenomenon of capacitors is the self-resonant frequency. Above this frequency, the capacitor ceases to look like a capacitor, and looks more like an inductor, although it still blocks DC voltages.

The instance where high frequency effects of capacitors come into play for PLL design is in power supply decoupling and the high frequency input pin. For the power supply pins, it is good practice to put a small capacitor for higher frequencies and a large capacitor for smaller frequencies. As for the high frequency input pin, a series capacitor is usually required to block DC voltages at this pin. Because the input impedance of a PLL is typically capacitive, it might actually be beneficial to exceed the self-resonant frequency of the capacitor in order to better match this impedance. If the capacitor chosen for this purpose is too small, then impedance matching problems can result. As a rough rule, 100 pF is a good value for frequencies of 500 MHz up to about 2 GHz, and beyond this, it might make sense to decrease this to a lower value.

Resistors

The real-world resistor has an equivalent parallel capacitance (EPC) and equivalent series inductance (ESL). Although this is component specific, a rough rule of thumb is to assume that the EPC = 0.2 pF and ESL = 1 nH. One guideline to get from this model is not to believe high resistance values at high frequencies. For instance, a real 1 kΩ resistor at 2 GHz operation is probably going to look a lot different than an ideal resistor under these conditions.

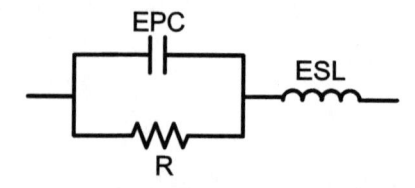

Figure 31.7 *High Frequency Model for a Resistor*

Conclusion

Although impedance matching networks are often unnecessary for matching the PLL to the VCO, there are enough situations where they are needed. Actually, what is really more critical is that the PLL input impedance be matched to the characteristic impedance of the PCB trace. When the trace length between the VCO and PLL approaches one-tenth of a wavelength, the trace is considered long and undesired transmission line effects can result. If there is plenty of VCO power to spare, the resistive pad serves as an economical and process-resistant solution. Otherwise, if the PLL is grossly mismatched to the VCO, the approach with inductors and capacitors can provide a good match. When using any sort of matching network, it is important to put this network as close to the PLL as possible.

References

Online Microstrip Impedance Calculator Tool
http://www.emclab.umr.edu/pcbtlc/microstrip.html

Danzer, Paul (editor) *The ARRL Handbook (Chapter 19)* The American Radio Relay
    League. 1997

I had useful conversations with Thomas Mathews regarding real-world component behavior.

Chapter 32   Other PLL Design and Performance Issues

Introduction

This is a collection of small topics that have not been addressed in other chapters. Included topics are N counter determination, the relationship between phase margin and peaking, and counter sensitivity.

N Counter Determination

N Value Determination for a Fixed Output Frequency PLL

In the case that the output frequency of the PLL is to be fixed, the choice of a comparison frequency may not be so obvious. The comparison frequency should always be chosen as large as possible.   Recall the relationship between comparison frequency and output frequency:

$$Fout = \left(\frac{N}{R}\right) \bullet Xtal \qquad (32.1)$$

It therefore follows that:

$$\frac{N}{R} = \frac{Fout}{Xtal} \qquad (32.2)$$

Since the output frequency and crystal frequency are both known quantities, the right hand side of this equation is known and can be reduced to a lowest terms fraction. Once this lowest terms fraction is known, the numerator is the N value and the denominator is the R value. If this solution results in illegal N divider ratios, or comparison frequencies that are higher than the phase detector can operate at, then double the N and R values. If there are still problems, then triple them. Keep increasing these quantities until there are no illegal divide ratios and the comparison frequency is within the specification of the part. In the case where there is freedom to choose the crystal frequency, it is best to choose it so that it has a lot of common factors with the output frequency so that the N value is as small as possible.

N Value to Design for When the Output Frequency is a Range

It has already been shown that the loop bandwidth is proportional to the loop gain, K, which is in turn inversely proportional to the N counter value. It therefore follows that designing the N value for the geometric mean of the minimum and maximum values minimizes the variation of the loop bandwidth of the PLL from the value for which it was designed. In summary, design for:

$$N = \sqrt{N\min \bullet N\max} \qquad (32.3)$$

The only exception would be when the N value or loop gain varies considerably. In this case, design somewhere between this recommended N value and the lowest N value in order to preserve loop stability.

Phase Margin, Stability, and Peaking

The phase margin is related to the stability of the system and a higher phase margin implies more stability. This can be seen by looking at the roots of the closed loop transfer function and tracking how negative the real parts of these roots are. On the spectrum analyzer, if the phase margin is very low, then the loop filter response will show a peaking. Recall that the closed loop transfer function is of the form:

$$CL(s) = \frac{G(s)}{1 + G(s)/N} \qquad (32.4)$$

Of special interest is at the point where the magnitude of $G(s)/N = 1$. The frequency where this occurs is, by definition, the loop bandwidth. The phase of $G(s)/N$ evaluated at the loop bandwidth is also of interest. If this phase is 180 degrees, then the transfer function would have an infinite value and would be unstable. If the phase were zero degrees, then there would be a minimal amount of peaking and maximum stability. Phase margin is therefore defined as the amount of margin on the phase which would be 180 degrees minus the phase of $G(j \bullet \omega c)/N$. In practice, loop filters with less than 20 degrees phase margin are likely to show instability problems and filters above 80 degrees phase margin have yield components that unrealistic because they are too large, or are negative. Second order formulas for lock time calculations imply that the lower phase margins imply a faster lock time, but when all poles and zeros are considered, it turns out that the optimal lock time is for phase margins in the 45 to 50 degree range.

On the Pitfalls of Sensitivity

Sensitivity is a feature of real world PLLs. The N counter will actually miscount if too little or too much power is applied to the high frequency input. The limits on these power levels are referred to as the sensitivity. The PLL sensitivity changes as a function of frequency. At the higher frequencies, the curve degrades because the of process limitations, and at the lower frequencies, the curve can also degrade because of problems with the counters making thresholding decisions (the edge rate of the signal is too slow). At the lower frequencies, this limitation can sometimes be addressed by running a square wave instead of a sine wave into the high frequency input of the PLL. Sensitivity can also change from part to part, over voltage, or over temperature. When the power level of the high frequency input approaches sensitivity limits, this can introduce spurs and degradation in phase noise. When the power level gets even closer to this limit, or exceeds it, then the PLL loses lock.

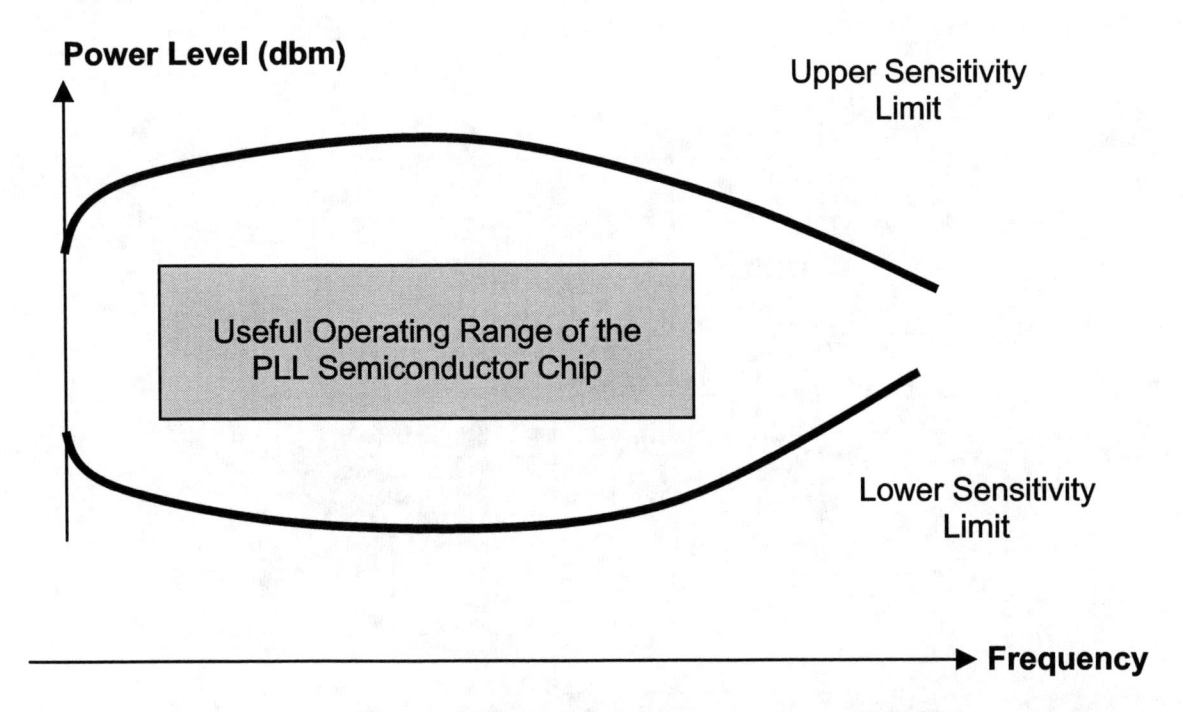

Figure 32.1 *Typical Sensitivity Curve for a PLL*

The sensitivity curve applies to both the desired signal from the VCO and all of its harmonics. VCO harmonics can especially be troublesome when a part designed for a very high operating frequency is used at a very low operating frequency. Unexpected sensitivity problems can also be agitated by poor matching between the VCO output and the high frequency input of the PLL.

Although sensitivity issues are most common with the N counter, because it usually involves the higher frequency input, these same concepts apply to the R counter as well. In order to for the sensitivity of the PLL to be tested in production, it is necessary to have access to the R and N counters. These test modes are also an excellent way of diagnosing and debugging sensitivity problems. Sensitivity related problems also tend to show a strong dependence on the Vcc voltage and temperature. If poor impedance matching is causing the sensitivity problem, then sometimes pressing one's finger on the part will temporarily make the problem go away. This is because the input impedance of the part is being impacted.

Sensitivity problems with either the N or R can cause spurs to appear, increase phase noise, or cause the PLL to tune to a different frequency than it is programmed to. In more severe cases, they can cause the PLL to steer the VCO to one of the power supply rails. N counter sensitivity problems usually cause the VCO to go higher than it should. R counter sensitivity problems usually cause the PLL to tune lower than it should. In either case, the VCO output is typically very noisy. Figure 32.2 shows a PLL locking much lower than it is programmed to lock due to an R counter sensitivity problem. It is also possible for the N counter to track a higher harmonic of the VCO signal, which causes the PLL to tune the VCO lower than it should. This problem is most common when parts are operated at frequencies much lower than they are designed to run at. One should be aware that it is possible to be operating within the datasheet specifications for sensitivity with a few dB of margin, and still have degraded phase noise as a result of a sensitivity problem. This is because the datasheet specification for sensitivity is a measurement of when the counters actually miscount, not when they become noisy.

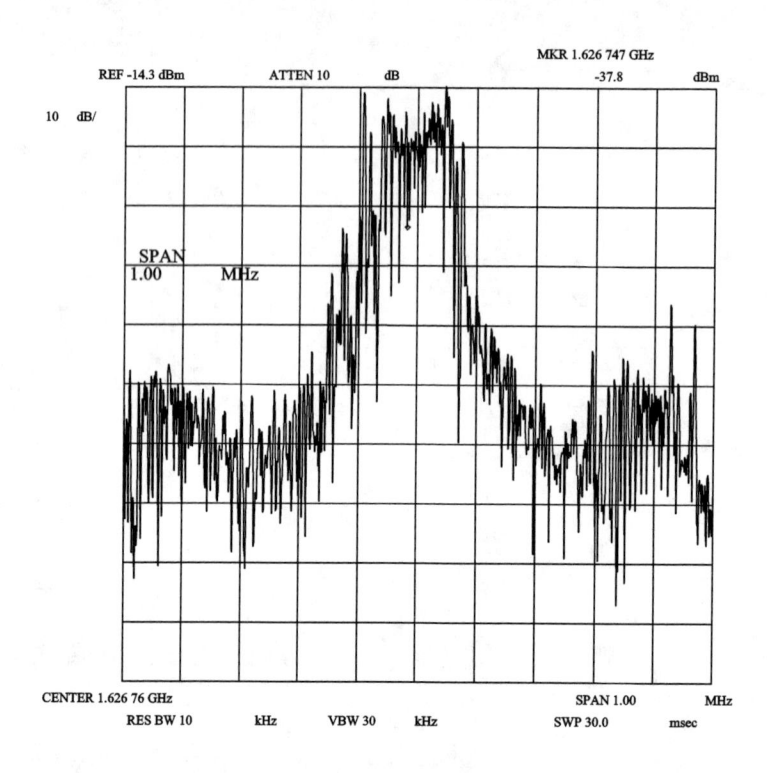

Figure 32.2 *PLL Locking to Wrong Frequency Due to R Counter Sensitivity Problem*

Impact of Using Dividers and Multipliers with a PLL

Dividers and Multipliers used between the VCO and the PLL Frequency Input Pin
In this application, there is a divider between the VCO and the PLL frequency input pin. The most common reason for doing this is that the VCO frequency exceeds the maximum frequency specification for the PLL. For this type of application, the N counter value is simply multiplied by the divider value. For instance, if a divide by two divider was used, then the effective N counter value would just be twice that used in the PLL chip. All the design equations and theory would still hold.

It would be awkward to use a multiplier in this type of application, but if one was used, the impact would simply be like dividing the N counter value by the multiplier value.

Dividers and Multipliers used After the VCO
For this type application, the divider or multiplier is placed after the complete PLL system. For the sake of simplicity, it will be assumed that the device is a divide by two divider. The first thing that one should realize that complete spectrum is 6 dB lower after the divider. One interesting thing is that the spur offset frequency is not impacted. Figure 32.3 illustrates that the result of dividing a signal with a 60 kHz spur by two still only has 60 kHz, no 30 kHz spurs.

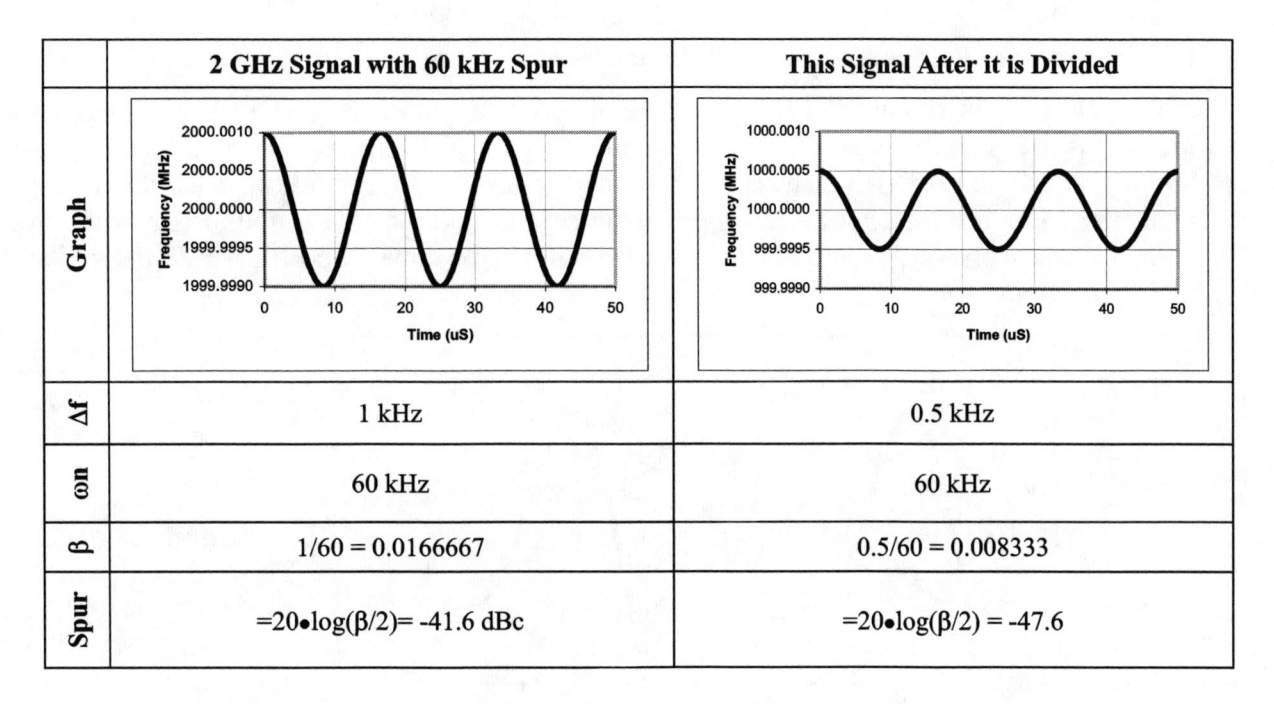

	2 GHz Signal with 60 kHz Spur	This Signal After it is Divided
Graph	*(frequency vs. time plot, 1999.9990–2000.0010 MHz, 0–50 μs)*	*(frequency vs. time plot, 999.9990–1000.0010 MHz, 0–50 μs)*
Δf	1 kHz	0.5 kHz
ωn	60 kHz	60 kHz
β	1/60 = 0.0166667	0.5/60 = 0.008333
Spur	$=20 \cdot \log(\beta/2) = -41.6$ dBc	$=20 \cdot \log(\beta/2) = -47.6$

Figure 32.3 *Impact of a Divider on a Spur*

The impact of removing the first spur (and all other odd spurs) is very significant, since the other spurs tend to be much easier to filter. The level of the even spurs depends on the type of PLL used. For an integer PLL, these spurs before the divider would be 6 dB higher, but then 6 dB lower after the divider. So the net effect is that they would be the same level. For a fractional PLL, the in-band spur does not increase with the N counter value, so it would not be increased by doubling the frequency of the VCO. However, after this is divided by two, then the net effect is that all the even fractional spurs would be 6 dB lower and all the odd fractional spurs would be completely gone. So the benefits of using a divider to lower spurs should be clear. If the value of the divider was larger, then the impact would be larger because it is a 20•log relationship.

As for phase noise, this also depends on the type of PLL used. For an integer PLL, the phase noise before the divider would be 3 dB higher due to the net effect of doubling the comparison frequency and leaving the N counter value the same. However, after the divider is considered, the net effect would be a 3 dB improvement. For a fractional PLL, this is also possible, but many times, the comparison frequency is already limited by the maximum phase detector rate or illegal divide ratios.

So the benefits of using dividers should be clear. Larger dividers have larger theoretical benefits. There are drawbacks, such as increased current consumption and reduced tuning range. Multipliers work just the opposite. If a multiplier is placed after the PLL, then the PLL itself needs to have half the channel spacing, which brings in more spurs.

PLL Accidentally Locking to VCO Harmonics

All VCOs put out harmonics. If the harmonic levels are too high, the PLL may lock to them instead of the intended signal. But what is too high? The theoretical result can be found by looking at the sum of two sine waves and inspecting what amplitude of a harmonic causes a miscount. For instance, when considering the second harmonic, it is found that if the voltage level is exactly one-half of the fundamental, which is 6 dB down, the PLL would theoretically be just about to miscount.

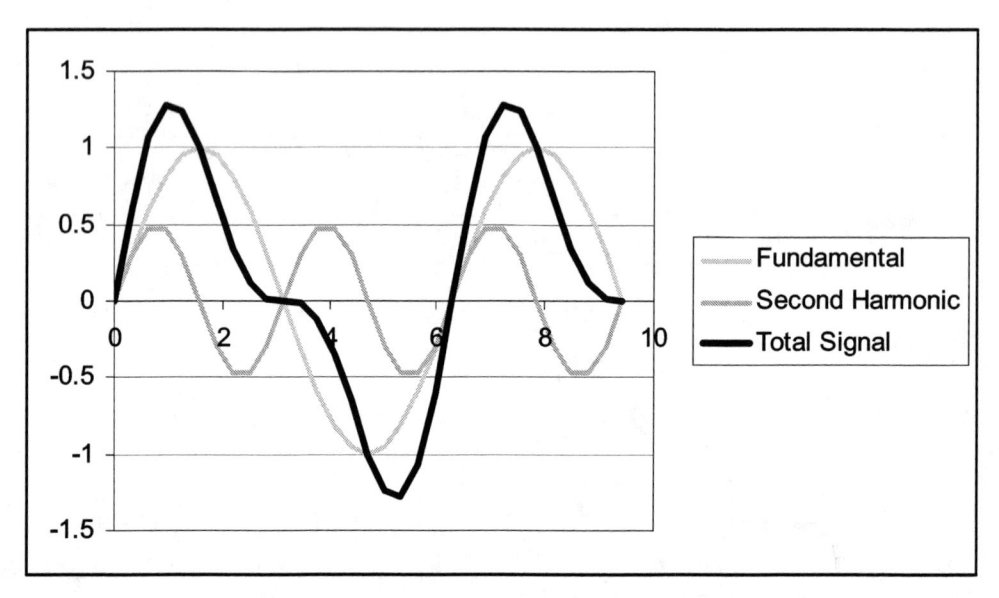

Figure 32.4 *Second Harmonic Illustration*

Assuming that these signals are in phase, the maximum tolerable harmonic for the higher order harmonics can also be calculated. Note that the even harmonics are much more of a problem than the odd harmonics. In fact, if the odd harmonics are in just the right mixture to make a square wave, the sensitivity is actually theoretically improved.

Harmonic	Maximum Tolerable Level *dBc*
2nd	-6.0
3rd	0.0
4th	-12.0
5th	-1.9
6th	-15.6
7th	-4.3
8th	-5.3
9th	-6.2
10th	-7.0

Table 32.1 *Theoretical Maximum Tolerable Harmonics*

The table above gives a theoretical maximum for the harmonic levels. However, as the harmonics approach the maximum tolerable levels, it becomes easier for any noise riding on the signal to cause the counters to miscount. There is also the fact that the PLL sensitivity varies as a function of frequency and will probably be different for the fundamental and harmonic. A LMX2326 PLL was tested with two signal generators. One signal generator simulated the fundamental frequency, where the second signal generator was used to simulate the second harmonic. It was found that the closer that the main signal was to the sensitivity limits, the more sensitive it was to the second harmonic. The sensitivity numbers used for the calculations here are actual measured data, not the datasheet limits, which tend to be much more conservative to accommodate for voltage, temperature, and process. The Normalized Harmonic is calculated by finding the harmonic level as normal, but then adjusting this:

$$\textit{Normalized Harmonic} = \quad (32.5)$$
$$\textit{(Harmonic Signal Strength - Fundamental Signal Strength)}$$
$$+ \ \textit{(Sensitivity to Fundamental Signal - Sensitivity to Harmonic Signal)}$$

Sensitivity Margin	Max Tolerable Normalized Harmonic
1 dB	-12 dBc
5 dB	-5 dBc
10 dB	-2 dBc
20 dB	0 dBc

Table 32.2 *Maximum Tolerable Normalized Second Harmonic*

For instance, consider an application where the user is operating at 400 MHz output with a +2.0 dBm signal. Further suppose that the sensitivity limit on this part is measured to be –8 dBm at 400 MHz and –20 dBm at 800 MHz. This means that this application has 10 dB margin on the sensitivity and can tolerate a normalized harmonic of –2 dBc, which translates to a harmonic level of –12 dBc after the sensitivity difference is considered. However, this does not have any margin. So if one adds in 5 dB margin, that works out to –17 dBc. Note that this table is empirical and not exact, but does serve as a rough guideline as to what harmonic levels are tolerable.

Note that there is a discrepancy between the theoretical results and the measured results. Theoretically, a second harmonic of greater than –6 dBc would cause a miscount, yet the measured results show 0 dBc is tolerable before considering sensitivity. The true answer probably lies somewhere between the theoretical and measured results, but is not critical to be exact because the whole goal is to stay away from these marginal designs.

Common Problems and Debugging Techniques for PLLs

Things often do not work the same way in practice as they do on paper. Or for that matter, the first PLL design often does not work at all. This section gives three common steps to get a PLL design up and working.

Step 1: *Confirm that the PLL is Responding to Commands Sent*

This is actually one of the most common problems. If a PC is being used to drive the PLL programming, this step is greatly simplified. Usually there is a bit that can be used to power the PLL up and down. If this bit is toggled, the current consumption should change, provided there is sufficient resolution on the current meter. Also, the high frequency input pins, and the crystal input pin usually have a DC bias level when the part is powered up (typically 1.6 volts), and zero volts when the part is powered down. If there is no power down bit, then sometimes there are I/O pins that can be toggled and observed. If none of these things can be done, proceed to *Step 2*. If there is a problem with this step, there are several possible causes.

If a PC is being used, the parallel port may not be working, or there could be a conflict. The operating instructions for the CodeLoader 2 software at wireless.national.com has a lot of information on things that could go wrong with the parallel port. There could be problems with the voltage levels also. Low pass filters put on the CLOCK and DATA lines can also cause programming problems. Another possibility is that the PLL is actually being programmed, but is powered down due to the state of some bit or some pin. Some PLLs will also not program if the crystal reference or VCO is not connected.

Step 2: *Confirm that the Carrier Frequency Can Be Changed*

The next step is to confirm that the carrier frequency can be moved. This can be done by toggling the phase detector polarity bit or programming the counters. Another technique is to program the N counter to zero and it's maximum value to see if the carrier will move.

Besides the reasons presented in *Step 1*, there are several things that could cause this problem. One common problem is that the PCB board actually accommodates a higher loop filter order than is needed, and 0 Ω resistors are not placed for the higher order resistors. Another possibility is that the loop filter is shorted to ground. This can be checked with a ohmmeter or it should also be apparent from the current consumption.

Sometimes, it is the case that the VCO frequency actually can be changed and the user makes some sort of mistake. For instance, if the span used on the spectrum analyzer is too large relative to the VCO tuning range, then it could appear that the PLL frequency is not changing, when it actually is. Many spectrum analyzers show a frequency spike at 0 Hz, which can sometimes also be mistaken for a signal. Yet another mistake sometimes done is to attempt to tune the VCO beyond its frequency range. In this case, it just stays at the frequency rails.

Step 3: ***In the Case of a PLL Carrier that Does React, but Shows Peaking, Instability, or Lock to the Wrong Frequency...***

Peaking and Instability

One possible problem is for the loop filter components to be wrong. One quick way to diagnose any loop filter issue is to observe the impact of reducing the loop gain, *K*. Also, if a loop filter is not very stable, this also shows up as an excessive lock time with a lot of ringing. This can be done by reducing the charge pump current or increasing the *N* counter value. A common mistake is to accidentally switch the capacitors *C1* and *C2* in the loop filter. Usually, the PLL will lock in this case, but there will be severe peaking. Another thing that can cause peaking or instability is when the VCO input capacitance is large compared to the capacitors it adds in parallel with. Yet a third common problem is for the VCO gain or charge pump gain to be off, which can cause peaking and instability. Aside from issues with the loop filter, sensitivity issues can cause a "Christmas Tree" spectrum which looks like instability.

Lock to the Wrong Frequency

The first thing to observe here is if the PLL locks clean or if there is a lot of noise. If there is a lot of noise, the cause could be sensitivity or harmonics. Both of these have already been discussed. One other mistake is to mistake one of the VCO harmonics for the actual carrier. If the PLL locks clean, this is more likely to be a programming error, or a attempt to program an illegal divide ratio.

Conclusion and Author's Parting Remarks

This chapter has addressed some of the issues not addressed in other chapters. The reader who has reached this point in this book should hopefully have an appreciation on how involved PLL design and simulation can be.

It was the aim of this book to tell the reader everything they wanted to know, and things they probably never cared to know about the designing and simulating a PLL frequency synthesizer. However, there are still many other topics that have been left out. The concepts presented in this book have come from a solid theoretical understanding backed with measured data and practical examples. All of the data in this book was gathered from various National Semiconductor Synthesizer chips, which include the *R* counter, *N* counter, charge pump, and phase-frequency detector.

Supplemental Information

Chapter 33   Glossary and Abbreviation List

Atten

The attenuation index, which is intended to give an idea of the spurious attenuation added by the components *R3* and *C3* in the loop filter of other loop filter design papers, but not this book. Also used in reference to the attenuation of a resistive pad in dB.

Bloomer

(slang) A very high spur that –30 dBc or higher and part of a collection of undesired spurs. If the spur is in-band, the spur needs to be –10 dBc or higher to be classified as a bloomer.

Channel and Channel Spacing

In many applications, a set of frequencies is to be generated that are evenly spaced apart. These frequencies to be generated are often referred to as channels and the spacing between these channels is often referred to as the channel spacing.

Charge Pump

Used in conjunction with the phase-frequency detector, this device outputs a current of constant amplitude, but variable polarity and duty cycle. It is usually modeled as a device that outputs a steady current of value equal to the time-averaged value of the output current.

Closed Loop Transfer Function , *CL(s)*

This is given by $\dfrac{G(s)}{1+G(s)\bullet H}$, where $H=\dfrac{1}{N}$ and *G(s)* is the Open Loop Transfer Function

Comparison Frequency, *Fcomp*

The crystal reference frequency divided by the *R* counter value. This is also sometimes called the reference frequency.

Continuous Time Approximation

This is where the discrete current pulses of the charge pump are modeled as a continuous current with magnitude equal to the time-averaged value of the current pulses.

Control Voltage , *Vtune*

The voltage that controls the frequency output of a VCO.

Crystal Reference, *Xtal*

A stable and accurate frequency that is used for a reference.

Damping Factor, ζ

For a second order transient response, this determines the shape of the exponential envelope that multiplies the frequency ringing.

Dead Zone

This is a property of the phase frequency detector caused by component delays. Since the components making up the PFD have a non-zero delay time, this causes the phase detector to be insensitive to very small phase errors.

Dead Zone Elimination Circuitry

This circuitry can be added to the phase detector to avoid having it operating in the dead zone. This usually works by causing the charge pump to always come on for some minimum amount of time.

Delta Sigma PLL

A fractional PLL that achieves fractional N values by alternating the N counter value between two or more values. Usually, the case of two values is considered a trivial case.

Fractional Modulus, *FDEN*

The fractional denominator used for in the fractional word in a fractional PLL.

Fractional *N* PLL

A PLL in which the *N* divider value can be a fraction.

Fractional Spur

Spurs that occur in a fractional N PLL at multiples of the comparison frequency divided by the fractional modulus that are caused by the PLL.

Frequency Jump, *Fj*

When discussing the transient response of the PLL, this refers to the frequency difference between the frequency the PLL is initially at, and the final target frequency.

Frequency Synthesizer

This is a PLL that has a high frequency divider (N divider), which can be used to synthesize a wide variety of signals.

Frequency Tolerance, *tol*

In regards to calculating or measuring lock time, this is the frequency error that is acceptable. If the frequency error is less than the frequency tolerance, the PLL is said to be in lock. Typical values for this are 500 Hz or 1 kHz.

Gamma Optimization Parameter, γ

A loop filter parameter that has some impact on the lock time. Usually chosen roughly close to one, but not exactly.

$$\gamma = \frac{\omega c^2 \cdot T2 \cdot A1}{A0}$$

G(s)

This represents the loop filter impedance multiplied by the VCO gain and charge pump gain, divided by s.

$$G(s) = \frac{K\phi \cdot Kvco}{s} \cdot Z(s)$$

Kvco

The gain of the VCO expressed in MHz/V.

Kϕ

This is the gain of the charge pump expressed in mA/(2π radians)

Locked PLL

A PLL such that the output frequency divided by *N* is equal to the comparison frequency within acceptable tolerances.

Lock Time

The time it takes for a PLL to switch from an initial frequency to a final frequency for a given frequency jump to within a given tolerance.

Loop Bandwidth, ωc or Fc

The frequency at which the magnitude of the open loop transfer function is equal to 1. ωc is the loop bandwidth in radians and Fc is the loop bandwidth in Hz.

Loop Filter

A low pass filter that takes the output currents of the charge pump and turns them into a voltage, used as the tuning voltage for the VCO. *Z(s)* is often used to represent the impedance of this function. Although not perfectly accurate, some like to view the loop filter as an integrator.

Loop Gain Constant

This is an intermediate calculation that is used to derive many results.

$$K = \frac{K\phi \bullet Kvco}{N}$$

Modulation Domain Analyzer

A piece of RF equipment that displays the frequency vs. time of an input signal.

Modulation Index, β

This is in reference to a sinusoidally modulated RF signal. The formula is given below, where F(t) stands for the frequency of the signal.

$$F(t) = const. + F_{dev} \bullet cos(\omega_m \bullet t)$$

$$\beta = \frac{F_{dev}}{\omega_m}$$

N Divider

A divider that divides the high frequency (and phase) output by a factor of N.

Natural Frequency, ωn

For a second order transient response, this is the frequency of the ringing of the frequency response.

Open Loop Transfer Function, $G(s)$

The transfer function which is obtained by taking the product of the VCO Gain, Charge Pump Gain and Loop Filter Impedance divided by N.

$$G(s) = \frac{K\phi \cdot Kvco \cdot Z(s)}{N \cdot s}$$

Overshoot

For the second order transient response, this is the amount that the target frequency is initially exceeded before it finally settles in to the proper frequency

Phase Detector

A device that produces an output signal that is proportional to the phase difference of its two inputs.

Phase-Frequency Detector, PFD

Very similar to a phase detector, but it also produces an output signal that is proportional to the frequency error as well.

Phase-Locked Loop, PLL

A circuit that uses feedback control to produce an output frequency from a fixed crystal reference frequency. Note that a PLL does not necessarily have an *N* divider. In the case that it does, it is referred to as a frequency synthesizer, which is the subject of this book.

Phase Margin, ϕ

180 degrees minus phase of the open loop transfer function at the loop bandwidth. Loop filters are typically designed for a phase margin between 30 and 70 degrees. Simulations show that around 48 degrees yields the fastest lock time. The formula is given below:

$$\phi = 180 - \angle C(j \cdot \omega c)$$

Phase Noise

This is noise on the output phase of the PLL. Since phase and frequency are related, it is visible on a spectrum analyzer. Within the loop bandwidth, the PLL is the dominant noise source. The metric used is dBc/Hz (decibel relative to the carrier per Hz). This is typically normalized to a 1 Hz bandwidth by subtracting 10*(Resolution Bandwidth) of the spectrum analyzer.

Phase Noise Floor

This is the phase noise minus $20 \bullet log(N)$. Note that this is generally not a constant because it tends to be dominated by the charge pump, which gets noisier at higher comparison frequencies.

Prescaler

Frequency dividers included as part of the N divider used to divide the high frequency VCO signal down to a lower frequency.

R Divider

A divider that divides the crystal reference frequency (and phase) by a factor of R.

Reference Spurs

Undesired frequency spikes on the output of the PLL caused by leakage currents and mismatch of the charge pump that FM modulate the VCO tuning voltage.

Resolution Bandwidth, *RBW*

See definition for Spectrum Analyzer.

Sensitivity

Power limitations to the high frequency input of the PLL chip (from the VCO). At these limits, the counters start miscounting the frequency and do not divide correctly.

Smith Chart

A chart that shows how the impedance of a device varies over frequency.

Spectrum Analyzer

A piece of RF equipment that displays the power vs. frequency for an input signal. This piece of equipment works by taking a frequency ramp function and mixing it with the input frequency signal. The output of the mixer is filtered with a bandpass filter, which has a bandwidth equal to the resolution bandwidth. The narrower the bandwidth of this filter, the less noise that is let through.

Spurious Attenuation

This refers to the degree to which the loop filter attenuates the reference spurs. This can be seen in the closed loop transfer function.

Spur Gain, *SG*

This refers to the magnitude of the open loop transfer function evaluated at the comparison frequency. This gives a good indication of how the reference spurs of two loop filters compare.

T31 Ratio

This is the ratio of the poles of a third order loop filter. If this ratio is 0, then this is actually a second order filter. If this ratio is 1, then this turns out to be the value for this parameter that yields the lowest reference spurs.

T41 Ratio

This is the ratio of the poles *T4* to the pole *T1* in a fourth order filter. If this ratio is zero, then the loop filter is third order or less.

T43 Ratio

This is the ratio of the pole *T4* to the pole *T3*. A rough rule of thumb is to choose this no larger than the *T31* ratio.

Temperature Compensated Crystal Oscillator, TCXO

A crystal that is temperature compensated for improved frequency accuracy

Varactor Diode

This is a diode inside a VCO that is reverse biased. As the tuning voltage to the VCO changes,

it varies the junction capacitance of this diode, which in turn varies the VCO voltage.

Voltage Controlled Oscillator, VCO

A device that produces an output frequency that is dependent on an input (Control) voltage.

Abbreviation List

Loop Filter Parameters

A0, A1, A2, A3	Loop Filter Coefficients
C1, C2, C3, C4	Loop filter capacitor values
CL(s)	Closed loop PLL transfer function
f	Frequency of interest in Hz
Fc	Loop bandwidth in kHz
Fcomp	Comparison frequency
FDEN	Fractional denominator or fractional modulus
Fj	Frequency jump for lock time
FNUM	Fractional Numerator
Fout	VCO output frequency
fn	VCO frequency divided by N
fr	XTAL frequency divided by R
Fspur	Spur offset frequency
G(s)	Loop filter transfer function
H	PLL feedback, which is 1/N
i, j	The complex number $\sqrt{-1}$
k	Fractional spur order
K	Loop gain constant.
Kϕ	Charge pump gain in mA/(2π radians)
Kvco	VCO gain in MHz/V
M	Loop bandwidth multiplier for Fastlock
N	The *N* counter Value
PFD	Phase/Frequency Detector
PLL	Phase-Locked Loop
r	Ratio of the spur frequency to the loop bandwidth
R	The *R* counter Value
R2, R3, R4	Loop filter resistor values
s	Laplace transform variable = $2\pi \bullet f \bullet j$
T2	The zero in the loop filter transfer function
T1, T3, T4	The poles in the loop filter transfer function

T31	The ratio of the pole *T3* to the pole *T1*
T41	The ratio of the pole *T4* to the pole *T1*
T43	The ratio of the pole *T4* to the pole *T3*
tol	Frequency tolerance for lock time
Vcc	The main power supply voltage
Vdo	The output voltage of the PLL charge pump
VCO	Voltage Controlled Oscillator
Vpp	The power supply voltage for the PLL charge pump
XTAL	Crystal Reference or Crystal Reference Frequency
Z(s)	Loop filter impedance

Greek Symbols

β	The modulation index
ϕ	The phase margin
ϕ_r	The XTAL phase divided by R
ϕ_n	The VCO phase divided by N
ω	The frequency of interest in radians
ω_c	The loop bandwidth in radians
ω_n	Natural Frequency
ζ	Damping Factor
γ	Gamma Optimization Parameter

Chapter 34   References

Best, Roland E.,   *Phase-Locked Loop Theory, Design, and Applications,* 3rd ed, McGraw-Hill, 1995

Danzer, Paul (editor) *The ARRL Handbook (Chapter 19)* The American Radio Relay League. 1997

Franklin, G., et. al., *Feedback Control of Dynamic Systems,* 3rd ed, Addison-Wesley, 1994

Gardner, F., *Charge Pump Phase-Lock Loops,* **IEEE Trans. Commun**. Vol COM-28, pp. 1849-1858, Nov. 1980

Gardner, F., *Phaselock Techniques,* 2nd ed., John Wiley & Sons, 1980

Keese, William O.   *An Analysis and Performance Evaluation for a Passive Filter Design technique for Charge Pump Phase-Locked Loops.*   AN-1001, National Semiconductor Wireless Databook

Lascari, Lance   *Accurate Phase Noise Prediction in PLL Synthesizers,* **Applied Microwave & Wireless,** Vol.12, No. 5, May 2000

Tranter, W.H. and R.E. Ziemer *Principles of Communications Systems, Modulation, and Noise,* 2nd ed, Houghton Mifflin Company, 1985

Weisstein, Eric   *CRC Concise Encyclopedia of Mathematics,* CRC Press 1998

Chapter 35   Useful Websites and Online RF Tools

http://www.anadigics.com/engineers/Receiver.html

Online receiver chain analysis tool for calculation of gain, noise figure, third order intercept point, and more.

http://www.emclab.umr.edu/pcbtlc/microstrip.html

This is an online microstrip impedance calculator that is useful in calculating the impedance of a PCB trace. It is very easy to use and also can be used to calculate the desired trace width in order to produce a desired impedance

http://www.geocities.com/szu_lan/

The author's personal website with both personal and professional information.

http://www.radioelectronicschool.com/raecourse.html

This page has many different lecture notes for a broad variety of electrical engineering topics.

http://tools.rfdude.com/

Lance Lascari's RF Tools Page. The Mathcad based PLL design worksheet is pretty good.

http://www-sci.lib.uci.edu/HSG/RefCalculators.html

Jim Martindale's calculators for everything you can think of.

http://www.treasure-troves.com

The "Rolls Royce" of mathematics online reference site on the web. There is also a corresponding book, which is excellent. Compiled by Eric Weisstein.

http://wireless.national.com

National Semiconductor's wireless portal site. It contains the EasyPLL program for PLL selection, design, and simulation. The EasyPLL program is largely based on this book. There is also analog university which contains self-paced coursework for PLLs complete with certificates of completion that can be earned. There is also programming software, evaluation boards, datasheets, and much more.